Trace Elements in the Environment: Biogeochemistry and Biotechnology

Trace Elements in the Environment: Biogeochemistry and Biotechnology

Gracie Price

CLANRYE
INTERNATIONAL
www.clanryeinternational.com

Clanrye International,
750 Third Avenue, 9ᵗʰ Floor,
New York, NY 10017, USA

ISBN: 978-1-64726-661-5

Cataloging-in-Publication Data

Trace elements in the environment : biogeochemistry and biotechnology / Gracie Price.
p. cm.
Includes bibliographical references and index.
ISBN 978-1-64726-661-5
1. Trace elements--Environmental aspects. 2. Trace elements.
3. Biogeochemistry. 4. Biotechnology. 5. Environmental chemistry.
I. Price, Gracie.
QH545.T7 T73 2023
572.515--dc23

For information on all Clanrye International publications
visit our website at www.clanryeinternational.com

Contents

Preface

A trace element refers to a type of chemical element whose concentration in the environment is very low. Different types of trace elements have been divided into two categories, namely, essential trace elements and non-essential trace elements. Natural processes and anthropogenic activities are responsible for the release of trace elements in the environment. Arsenic (As), antimony (Sb), cadmium (Cd), chromium (Cr), and lead (Pb) are some of the trace elements that deteriorate the quality and properties of soil. The biogeochemistry of trace elements from soils to rivers is studied for its impact on aquatic ecosystem toxicity. Microbes are crucial in the biogeochemical cycling of trace elements. The global biogeochemical cycle of trace elements is dependent on the speciation, mobility and bioavailability of microbes. This book provides a detailed explanation of the various trace elements found in the environment along with their biogeochemical and biotechnological aspects. It will provide comprehensive knowledge to the readers.

This book unites the global concepts and researches in an organized manner for a comprehensive understanding of the subject. It is a ripe text for all researchers, students, scientists or anyone else who is interested in acquiring a better knowledge of this dynamic field.

I extend my sincere thanks to the contributors for such eloquent research chapters. Finally, I thank my family for being a source of support and help.

Gracie Price

Anaerobic Digesters and Biogeochemistry

Eric D. van Hullebusch[1,2], Sepehr Shakeri Yekta[3],
Baris Calli[4] and Fernando G. Fermoso[5]*

[1]*Institut de Physique du Globe de Paris, Sorbonne Paris Cité, Université
Paris Diderot, UMR 7154, CNRS, F-75005 Paris, France*
[2]*Department of Environmental Engineering and Water Technology, IHE Delft
Institute for Water Education, Delft, The Netherlands*
[3]*Department of Thematic Studies-Environmental Change, Linköping
University, Linköping, Sweden*
[4]*Environmental Engineering Department, Marmara University, Istanbul,
Turkey*
[5]*Instituto de la Grasa (C.S.I.C.), Sevilla, Spain*
Corresponding author E-mail: vanhullebusch@ipgp.fr

ABSTRACT

This chapter provides an overview of the main biochemical transformations of
major elements, including carbon, nitrogen, sulfur, iron and phosphorus in
anaerobic digesters. Mineralization of organic matter during anaerobic digestion
processes results in the production of inorganic carbonate, ammonium, sulfide,
and phosphate species, which are involved in a complex network of chemical and
biological reactions through interaction with available macro and micro nutrients
as well as microbial processes with profound effects on the efficiency and
stability of the anaerobic digester performance. The interplay of iron, phosphorus
and sulfur cycles has recently attracted attention in the frame of research
developed for the recovery of phosphorus on one hand and in the frame of the

addition of iron minerals for enhancing anaerobic digester performance on the other hand. These research topics have led to the development of analytical and biogeochemical modelling tools to provide a better understanding of the fate of major elements in an anaerobic digester.

KEYWORDS: anaerobic digestion, biogeochemical cycles, biogeochemistry, interplay, major elements

1.1 INTRODUCTION

Anaerobic digestion (AD) is a well established technology for biological treatment of organic waste streams from industrial, municipal and agricultural activities and simultaneous production of renewable energy in form of bio-methane. Elemental composition of organic waste in terms of carbon, nitrogen, phosphorus, and sulfur content is the primary factor that determines the suitability of organic waste for application in AD processes, in which a balanced supply of nutritional

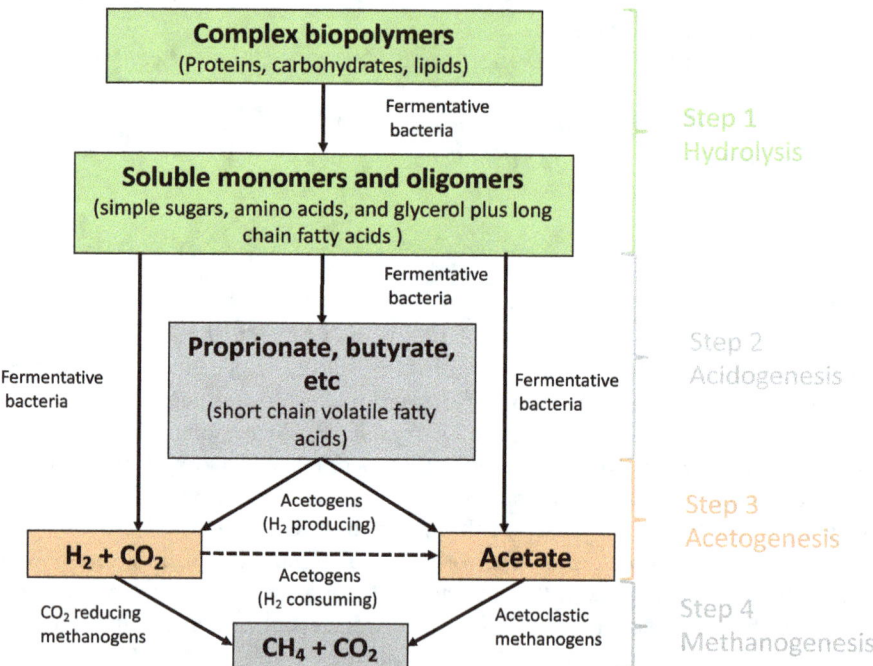

Figure 1.1 Schematic representation of the anaerobic degradation of organic matter AD divided in four steps: (1) hydrolysis, (2) acidogenesis, (3) acetogenesis, and (4) methanogenesis.

elements is crucial for establishment of well functioning microbial communities and efficient process performance. Organic wastes used in AD processes encompass a complex mixture of compounds with diverse chemical properties (Wellinger *et al.*, 2013). Accordingly, different microbial degradation pathways and chemical environments may be established in anaerobic digesters, depending on the overall composition of the influent organic wastes. Examples of the categories of compounds commonly present in organic wastes are biological solids (i.e. coagulated non-degraded solids from dead organisms and flocs produced via the interactions of organic matter), proteins, carbohydrates, lipids, microbial intermediate and end products as well as inorganic ions, halogen, alkali and alkaline earth metals. In particular, inorganic ions including carbonate, ammonium, phosphate, and sulfide are produced via mineralization of the organic compounds in the majority of AD processes with profound effects on the efficiency of the organic matter degradation reactions and process performance. This chapter addresses the major biological and chemical reactions, which control the biochemistry of major elements – carbon, nitrogen, phosphorus, sulfur and iron – in anaerobic digester environments. The interplay of some important elements such as iron, sulfur and phosphorus is also emphasized.

1.2 CARBON BIOGEOCHEMISTRY IN ANAEROBIC DIGESTERS

Carbon is found in both organic and inorganic forms in anaerobic digesters. Inorganic forms are dominated by dissolved CO_2 and carbonate ions produced via mineralization of organic carbon, as well as carbonate precipitates such as calcium carbonate ($CaCO_3$). Organic carbon forms encompass a complex pool of dissolved and particulate molecules, which are commonly expressed as biochemical groups of proteins, carbohydrates and lipids. The relative content of different organic groups largely depends on the origin of the organic wastes, for example, sewage sludge from wastewater treatment plants (WWTP), slaughterhouse waste, food waste, manure, crops or crop residues. Anaerobic digestion involves a solubilization of organic macromolecules with complex chemical structures into their organic sub-units, followed by conversion of the solubilized entities into carbon dioxide and methane (so called biogas) (van Lier *et al.*, 2008). The AD process involves four sequential degradation pathways (Figure 1.1), including: (i) hydrolysis, in which large molecules, such as carbohydrates, proteins and lipids, are converted in their monomers, that is, simple sugars, amino acids, glycerol, and by long chain fatty acids; (ii) acidogenesis where fermentable compounds (e.g. sugars and glycerol) are converted to volatile fatty acids; (iii) acetogenesis, in which acetate is synthesized from the oxidation of, for example, fatty acids by syntrophic bacteria (Angelidaki *et al.*, 2011), or from the utilization of H_2/CO_2 by homoacetogenic bacteria; and

(iv) methanogenesis, the final AD step in which simple compounds such as acetate and H_2/CO_2 are converted to biogas (Figure 1.1, modified from Paulo et al., 2015) (van Lier et al., 2008). Microbial growth and activities also contribute to incorporation of carbon in microbial cells and in compounds of microbial origins such as extracellular polymeric substances.

The fate of carbon in anaerobic digesters predominantly depends on the parameters which affect the degree and rate of organic matter degradation processes. During AD about 20–95% of the feedstock organic matter is degraded, depending on feedstock composition (Möller & Müller, 2012). Generally, characteristics of organic matter in the influent substrate in term of biodegradability together with operational conditions such as hydraulic retention time (HRT) and organic loading rate (OLR) determine the degradation and fate of carbon in anaerobic digesters (Wellinger et al., 2013). In cases where biogas production is in focus, the rate-limiting step is defined as the step that limits the conversion rate of organic compounds to biogas. It is well known that the rate-limiting step for organic substrates with complex chemical structures, such as cellulosic and hemicellulosic materials, is the hydrolysis step due to the constrained solubilization kinetics of recalcitrant carbon structures by hydrolytic enzymes, whereas methanogenesis is the rate-limiting step for easily-biodegradable substrates due to a slow growth and activity of methanogenic microorganisms. Furthermore, AD may suffer from low methane yield, instability and even process failure due to low concentration of trace elements, which are important micronutrients for microbial growth and activities, depending on the composition of feedstock (Dong et al., 2009; Jiang et al., 2012; Wei et al., 2014; Wellinger et al., 2013; Zhang et al., 2011). Thus, balancing the optimal nutrient composition as well as application of trace elements supplement may be required (Demirel & Scherer, 2011) especially when AD is operated on mono-substrate. The degree and kinetics of organic matter degradation is intertwined with biochemistry of major and trace elements, which will be discussed in detail in following sections.

The biochemistry of inorganic carbon (i.e. carbonate) is particularly important for regulation of pH and alkalinity in anaerobic digesters (Sun et al., 2016a, b). In general, dissolved concentrations of carbonates, ammonium, and organic acids (e.g. acetate, propionate, butyrate, etc.), produced as intermediate degradation products, control the pH and alkalinity of anaerobic digesters (Chen et al., 2015) according to following acid/base reactant concentrations:

$$CO_2 + H_2O \leftrightarrow HCO_3^- + H^+ \quad pKa = 6.4(25°C) \tag{1.1}$$

$$HCO_3^- + H^+ \leftrightarrow CO_3^{2-} + H^+ \quad pKa = 10.3(25°C) \tag{1.2}$$

$$NH_4^+ \leftrightarrow NH_3 + H^+ \quad pKa = 9.3(25°C) \tag{1.3}$$

$$CH_3COOH \leftrightarrow CH_3COO^- + H^+ \quad pKa = 4.75(25°C) \tag{1.4}$$

The pH of anaerobic digesters is considered as a critical parameter in steering metabolic pathways of the established microbial community, in which methanogens are often the groups most sensitive to pH changes in the digesters and the pH needs to be maintained within an optimum pH range (i.e. 6.5 and 8.0 based on optimum pH for growth of methanogens). Accordingly, buffering materials may be needed upon acidification of the reactor for pH control, when the kinetic balance between acid generation and consumption (i.e. microbial activity balance between the fast-growing acidogens and the slow-growing methanogens) is disturbed (Chen *et al.*, 2015). For instance, the addition of Na_2CO_3 and $NaHCO_3$ have been reported to significantly effect the stability of the anaerobic digesters (Dong *et al.*, 2009) by the neutralization of $2H^+$ and $1H^+$, respectively. A pH increase may also occur due to formation of ammonium carbonate (($NH_4)_2CO_3$) and the removal of CO_2 via biogas extraction as a result of the transformation of CO_3^{2-} and $2H^+$ to CO_2 and H_2O. Also pH may be increased by the supply of major cations (e.g. Ca^{2+} via the addition of $CaCO_3$; Chen *et al.*, 2015) because the electric-charge balance of the bioreactor solution has to be neutral, thus, decreasing the concentration of H^+ (Möller & Müller, 2012). In contrast, precipitation of carbonates (e.g. calcite $CaCO_3$) reduces digester pH (Möller & Müller, 2012) (Figure 1.2).

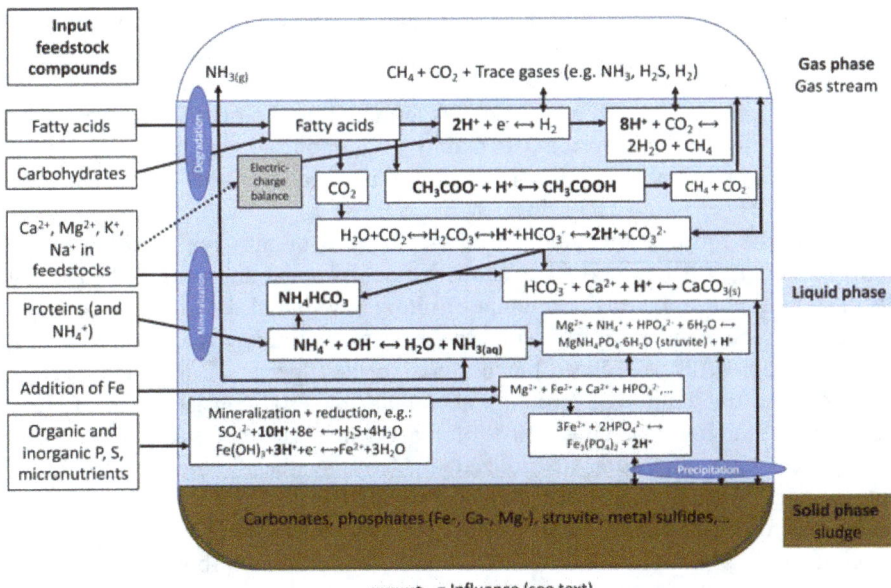

········▸ = Influence (see text)

Figure 1.2 Processes affecting pH value of anaerobic digesters (Modified from Möller & Müller, 2012).

1.3 NITROGEN BIOGEOCHEMISTRY IN ANAEROBIC DIGESTERS

Nitrogen is a key macronutrient that is essential to all organisms because it is a critical element of proteins, DNA and chlorophyll (Stüeken et al., 2016). It is the fourth most common element after carbon, oxygen and hydrogen in the biosphere. Dissimilar to other biologically crucial elements, nitrogen is a minor constituent of the Earth's crust (Ward, 2012). It is specifically abundant in the atmosphere as nitrogen gas (N_2). However, most organisms are not able to utilize it in N_2 form. Therefore, as a resource, nitrogen is scarce and often restricts primary productivity in the biosphere. It becomes available to primary producers (plants, algae, cyanobacteria, etc.) when transformed to ammonia (NH_3).

Nitrogen exists in various oxidation states from -3 to $+5$ and the oxidation state has a significant role in its cycle. The key nitrogen transformations are nitrogen fixation, ammonification, nitrification, denitrification and anammox. The conversion of nitrogen to different oxidation states relies heavily on the activities of different groups of organisms, such as bacteria, archaea and fungi (Bernhard, 2010).

During the decomposition of organic nitrogen compounds such as amino acids and nucleotides by various microorganisms in anaerobic digesters, inorganic nitrogen is released as ammonia. This process is called ammonification or mineralization, because the organic nitrogen is converted to an inorganic or mineral form. Deaminases play an important role in removal of amino group from proteins and amino acids and results in the release of ammonium (Ward 2012). The form of reduced ammonia is determined by the pH of the digester. At neutral pH, ammonia nitrogen exists mainly as ammonium ion (NH_4^+). However, at alkaline conditions the unionized free NH_3 becomes the dominant form (Equation 1.3).

NH_3 is the most readily utilized inorganic form of nitrogen, existing in the reduced state that is required for anabolic metabolism and an uncharged state that facilitates cellular uptake. The amount of nitrogen required for biomass synthesis in AD can be estimated by using the empirical formula of $C_5H_7O_2N$ that generally represents the anaerobic bacterial cells (Speece & McCarty, 1964). According to this empirical formula, the cellular material consists of about 12% nitrogen. Assuming that about 10% of the organic matter (as COD) removed during anaerobic digestion is utilized for cell synthesis, the corresponding nitrogen requirement is estimated as 1.2 g per 100 g of COD removed. Also additional processes involve traces of NH_3 volatilization ($<1\%$) and NH_4^+ precipitation as struvite ($MgNH_4PO_4$) and ammonium carbonate (NH_4CO_3) (Figure 1.3) (Möller & Müller, 2012).

On the other hand, high levels of NH_3 in anaerobic digesters inhibit microorganisms by inducing intracellular proton imbalance (Gallert et al., 1998).

Because equilibrium between NH_3 and NH_4^+ shifts towards higher concentration of NH_3, with increasing pH, ammonia inhibition is observed at high pH levels over 7.5–8 (Calli *et al.*, 2005). Besides, compared with mesophilic digesters, thermophilic ones are more vulnerable to ammonia inhibition given that NH_3 concentration also increases with temperature (Hansen *et al.*, 1998). In the literature, it is reported that the methane production generally starts to decrease when NH_3 exceeds 150 mg/L (Calli *et al.*, 2005).

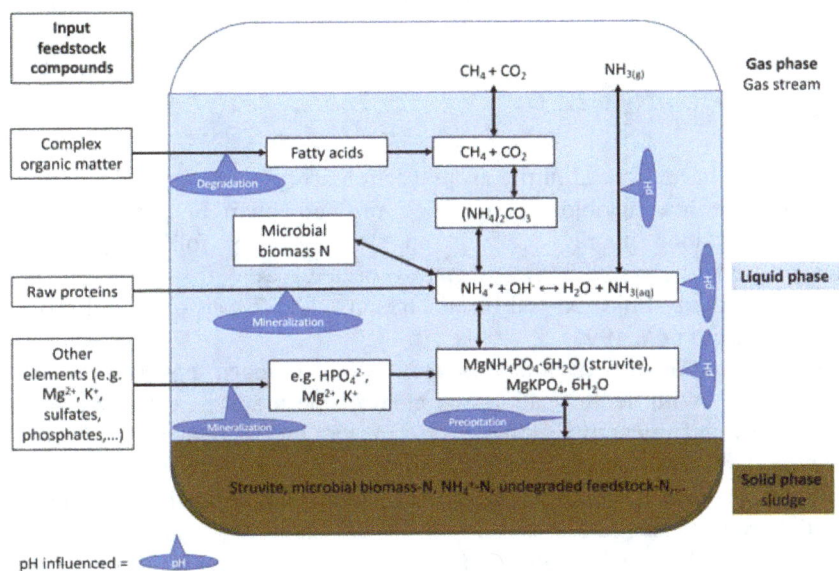

Figure 1.3 Nitrogen turnover in anaerobic digesters (Modified from Möller & Müller, 2012).

Among the microorganisms in anaerobic digesters, methanogens are the most vulnerable to ammonia inhibition (Kayhanian, 1994). It was reported that methanogens may be acclimated to total ammonium (NH_4^+ plus NH_3) at nitrogen concentrations over 5000 mg/L in manure digesters (Bayrakdar *et al.*, 2017). Acetoclastic methanogens are more sensitive to ammonia inhibition and that is why the methanogenic pathways shifts towards syntrophic acetate oxidation coupled to hydrogenotrophic methanogenesis at high amonia levels (Sun *et al.*, 2016a, b).

Nitrification is a critical process in the nitrogen cycle that is responsible for the conversion of ammonia to nitrite and then to nitrate. Under anaerobic conditions, it can occur only if alternative electron acceptors such as manganese oxides are available (Mortimer *et al.*, 2002). Therefore, nitrification is very rare in methanogenic digesters operating under strict anaerobic conditions.

Anaerobic ammonium oxidation using nitrite as an electron acceptor was realized in the mid-90 s, a long time after the discovery of aerobic ammonium oxidation, but it has an important role in nitrogen cycling in oxygen-limited anoxic environments (Mulder et al., 1995; van de Graaf et al., 1995).

In the anammox reaction, nitrite (NO_2^-) is first reduced to nitric oxide (NO) by nitrite reductase, and then NO reacts with ammonium (NH_4^+) to yield hydrazine (N_2H_4) which is a very strong reductant, by activity of the enzyme hydrazine hydrolase. N_2H_4 is then oxidized to N_2 by the enzyme hydrazine dehydrogenase (Jetten et al., 2009). A simplified reaction of anaerobic ammonium oxidation is given below.

$$NH_4^+ + NO_2^- \rightarrow N_2 + 2H_2O \qquad\qquad (1.5)$$

Anammox bacteria need nitrite as an electron acceptor. As the strict anaerobic conditions are not suitable for oxidation of ammonium to nitrite, anaerobic ammonia oxidation mainly occur in an anoxic reactor following nitritation process (kanders et al., 2018). Anammox bacteria were first identified in an anoxic denitrifying fluidized-bed reactor treating the effluent of a methanogenic reactor (Mulder et al., 1995).

The process in which nitrate is reduced to nitrogen gas (N_2) is called denitrification. With denitrification, the bioavailable nitrate is converted to inert nitrogen gas and released to atmosphere. In addition to final product nitrogen gas, there are some intermediate gaseous products such as nitrous oxide (N_2O) and nitrogen dioxide (NO_2). These intermediate denitrification products are potential greenhouse gasses and result in depletion of ozone layer and air pollution as well (Byrne & Goldblatt, 2014). The overall denitrification reaction is presented as follows:

$$2NO_3^- \rightarrow N_2 + 6H_2O \qquad\qquad (1.6)$$

Denitrification is an anaerobic process and generally occurs in oxygen-free environments. It is performed by a diverse group of prokaryotes, however, there are also some eukaryotes capable of denitrification (Kamp et al., 2015). Most denitrifiers are phylogenetically members of the *Proteobacteria* and physiologically facultative aerobes. Some denitrifying bacteria use ferric iron and some organic substances as electron acceptor in addition to nitrate (Strohm et al., 2007).

Denitrifiers have the ability to utilize a variety of fermentative and methanogenic substrates available in anaerobic digesters and they compete with the other microorganisms for the same substrates, such as glucose, volatile fatty acids and H_2. Accordingly, denitrification lowers the CH_4 yield per kg of organic matter in digesters (Akunna et al., 1992).

1.4 PHOSPHORUS BIOGEOCHEMISTRY IN ANAEROBIC DIGESTERS

One of the most important macronutrients for anaerobic microorganisms is phosphorus. In addition to the general role of providing the building blocks required for growth, phosphorus is also involved in many unique co-enzymes (Takashima & Speece, 1989). The phosphorus in an anaerobic digester consists mainly of organically bound/organic phosphates and metal-associated phosphates in the feedstock. For example, iron−phosphorus (FePs) compounds are the dominant metal-associated phosphates in sewage-sludge streams (Wilfert *et al.*, 2015). FePs found in WWTP can be either iron phosphate minerals or adsorption complexes which involve adsorption of orthophosphate ions onto iron oxides (Wilfert *et al.*, 2015).

During digestion the organically bound/organic phosphates are converted to soluble orthophosphate. At the most common operational pH, that is, 7.3, soluble orthophosphate ions are mainly distributed as 50% $H_2PO_4^-$ and 50% HPO_4^{2-}, with almost negligible amounts of H_3PO_4 and PO_4^{3-}. Besides soluble orthophosphate, in the soluble fraction of an anaerobic digester phosphorus is also present as dissolved organic phosphates and organic-phosphate complexes. In the solid fraction of an anaerobic digester, phosphorus can be present: as a mixture of phosphate precipitates and co-precipitate; as phosphate adsorbed to inorganic and organic surfaces; as organic phosphates; and/or intracellularly-accumulated inorganic phosphates (Güngör & Karthikeyan, 2005). Precipitation of ortho- and polyphosphate anions under conditions of anaerobic digestion is likely to occur as they are multiply charged negative species which easily react with multiply charged positive ions. The phosphate precipitates formed under conditions of anaerobic digestion depend primarily on the types and concentrations of metal ions in the digested sludge as well as on the pH of digesters (Van Rensburg *et al.*, 2003). Indeed as shown in Figure 1.4, pH may strongly influence the solubility of P and micronutrients. Raising the pH moves the chemical equilibrium toward the formation of phosphate ($HPO_4^{2-} \rightarrow PO_4^{3-}$) and subsequent precipitation of calcium phosphates and magnesium phosphates. Mineralization of N, P, and Mg combined with a substantial increase of the pH can enhance the formation and crystallization of struvite (Li *et al.*, 2019). Amorphous calcium phosphate, tricalcium phosphate or brushite ($CaHPO_4 \cdot 2H_2O$) are among the favorable calcium phosphates precipitates, which might transform to more stable forms over time. Phosphorus is also known to adsorb to and co-precipitate with inorganic compounds, particularly calcium carbonate under conditions of AD. Organic phosphates such as co-enzymes, nucleic acids, phosphorylated-carbohydrates and phospholipids, exist in the solid fraction of an anaerobic digester within biomass. Phosphate must always be considered to be in a transient state, as there is rapid cycling of organic phosphate to soluble orthophosphate and vice versa.

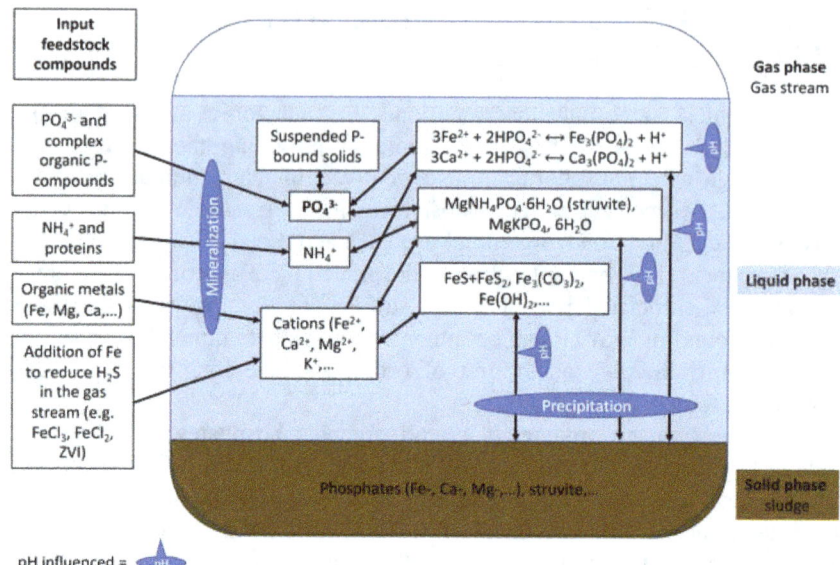

Figure 1.4 Phosphorus turnover in anaerobic digesters (Modified from Möller & Müller, 2012).

1.5 IRON BIOGEOCHEMISTRY IN ANAEROBIC DIGESTERS

Iron is a transition metal and its chemistry is quite complex (Kappler *et al.*, 2015). Iron can exist in various oxidation states (varying between -2 and $+6$), but the main iron species that occur naturally are either ferrous, Fe(II), or ferric iron, Fe(III), both being the most common oxidation states encountered in anaerobic digesters. The solubility of ferrous and ferric ions varies with pH and oxidation reduction potential (ORP) (Figure 1.5) (Wilfert *et al.*, 2015). The reactivity of ferrous and ferric ions depends on bulk conditions (temperature, pH, the nature of the complexing ligands, e.g. humic substances, etc.). Fe(II) is the more abundant redox species in anoxic environments whereas, in an oxygen-containing environment, iron may be readily oxidized from Fe(II) to Fe(III) state. Iron solubility and reactivity also strongly depend on pH, as spontaneous chemical oxidation of iron can be rapid at neutral pH whereas at low pH this abiotic oxidation occurs very slowly (Ilbert & Bonnefoy, 2013). Depending on the pH, the ferrous and ferric ions can generate various insoluble oxides, oxyhydroxides and hydroxides, together named iron oxides (Kappler *et al.*, 2015).

As previously mentioned, iron is known to react with oxygen in water or air moisture to form various insoluble iron oxide compounds described commonly as rust; there are many known iron oxides and oxyhydroxides reported in the literature (Kappler *et al.*, 2015). The most common iron oxides are ferrihydrite ($Fe_2O_3 \cdot 0.5H_2O$), hematite (Fe_2O_3) and magnetite (Fe_3O_4). Depending on the environment, iron not only complexes with oxygen ligands, but also with a lot of different compounds such as carbonate and sulfur by abiotic or biotic reactions (Kappler *et al.*, 2015). Such complexes can be found in anaerobic digesters environment such as siderite (iron carbonate: $FeCO_3$), ferrous sulfide (FeS), pyrite (FeS_2), vivianite ($Fe(II)_3[PO_4]_2 \cdot 8H_2O$), etc. (Kappler *et al.*, 2015; Wilfert *et al.*, 2015). This list is not exhaustive but allows an overview of the large diversity of iron mineral compounds available in anaerobic digesters (feedstock and digestate).

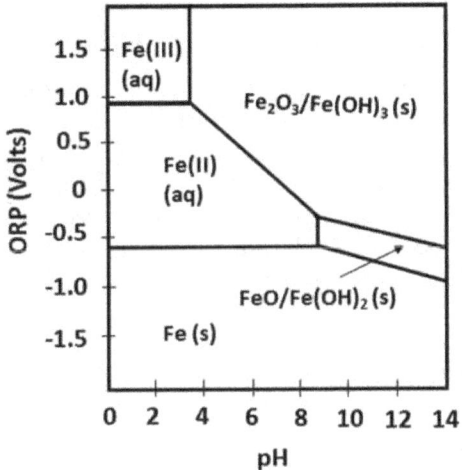

Figure 1.5 Simplified Pourbaix diagram showing the stable iron species under different conditions.

Iron recently became one of the most important cost-effective additives to improve anaerobic digestion performance due to its conductive properties. Indeed, iron is regularly added to AD systems as Fe(II) or Fe(III), and more recently the use of Zero Valent Iron (ZVI) has increased significantly (Romero-Güiza *et al.*, 2016; Wei *et al.*, 2018; Yang *et al.*, 2013). Reported iron minerals dosing advantage include: (i) its capacity to decrease the ORP of the anaerobic digestion media and therefore provide a more favorable environment for AD; (ii) its role as a cofactor of several key enzymatic activities, such as pyruvate-ferrodoxin oxidoreductase, which contains Fe–S clusters and plays a key role in fermentation; and (iii) its reaction with sulfide which allow to decrease the hydrogen sulfide level in the anaerobic digestion media and in biogas (Wei *et al.*, 2018). Different iron redox forms have been reported to stimulate anaerobic digestion. On the one hand, Fe(III) reduction is a favorable process to directly oxidize organics into simple compounds. The potential for ferric iron reduction with fermentable substrates, fermentation products, and complex organic matter as electron donors has been deeply investigated in sediments (Kappler et al., 2015; Lovley, 1997). However, Fe(III) bioreduction in anaerobic digesters may limit the conversion of organic matter to methane as microbial Fe(III) reduction is more thermodynamically favorable than methanogenesis. On the other hand, ZVI has been reported to accelerate the hydrolysis and fermentation step due to its role as an electron donor (Wei et al., 2018). When ZVI is added in the anaerobic digester, both chemical (Equation 1.7) and electrochemical corrosions (Equation 1.8) exist.

$$Fe + 2H_2O \rightarrow Fe(OH)_2 + H_2 \tag{1.7}$$

$$4Fe + SO_4^{2-} + 4H_2O \rightarrow 3Fe(OH)_2 + FeS + 2OH^- \tag{1.8}$$

Chemical corrosion results in the production of hydrogen. Hydrogen evolved from iron corrosion could serve as a substrate for hydrogenotrophic methanogens and homoacetogens consequently enhance CH_4 production. In terms of ORP, iron corrosion was reported to decrease ORP to a level suitable for AD (Feng et al., 2014). From Equation 1.8, one can directly see that sulfate may be electrochemically reduced by ZVI at the expense of sulfate reducing bacteria while sulfide precipitates as ferrous sulfide (Liu et al., 2015).

1.6 SULFUR BIOGEOCHEMISTRY IN ANAEROBIC DIGESTERS

Organic and inorganic sulfur (S) compounds are present in various organic wastes, used as substrates of AD processes. Microbial sulfate reduction and mineralization of organic S convert the S content of the organic wastes to sulfide, which at high levels perturbs the performance of anaerobic digesters. For example, sulfate favors the growth of sulfate-reducing microorganisms that compete with microbial groups involved in methane production for organic acids, decreasing the overall methane production yield of the AD processes (Raskin et al., 1996). Sulfide at high concentrations may as well impose microbial toxicity, inhibiting anaerobic degradation of organic matter and methane production (O'Flaherty et al., 1999). Presence of sulfide and consequent formation of metal-sulfide minerals influence the chemical speciation of metals by limiting the bioavailability of essential micronutrient metals, while attenuating the negative effects of toxic metals on microorganisms (Gonzalez-Estrella et al., 2015; Gustavsson et al., 2013). In addition, sulfide evolves into biogas, which necessitates the use of sulfide-removal techniques prior to application of biogas for heat and power generation. Despite the widely recognized challenges associated with AD of S-rich organic wastes, it is notable that S is an important nutrient for synthesis of cell compounds and is essential for efficient growth and activity of microorganisms (Merchant & Helmann, 2012).

Sulfur occurs in diverse biotic and abiotic forms in anaerobic digesters, depending on the substrate profile and operational condition of the AD processes. Chemical speciation studies have shown that S is present in both organic and inorganic forms with different valance states, ranging from reduced (valence state -2) to oxidized (valence state 6) S species (Hundal et al., 2000; Shakeri Yekta et al., 2014; Sommers et al., 1977; van Hullebusch et al., 2009). Sulfur speciation in anaerobic digesters is typically dominated by FeS, zero-valent S, and reduced organic S (i.e. organic sulfide, disulfide and thiol), whereas oxidized organic S as sulfate and sulfonate species occur to a minor extent. Supplementation of Fe for sulfide removal and application of Fe-rich substrates, such as primary and activated sewage sludge, are the main reason behind the formation of FeS in anaerobic digesters. However, presence of metals other than Fe may considerably contribute to the S speciation, depending on the relative concentrations of metals in the digesters.

Zero-valent S in anaerobic digesters may be related to formation of elemental S and/or pyrite. Elemental S can be formed through a series of reactions from aqueous sulfide as the main precursor at circumneutral pH, which involves a sequential formation of polysulfides and further acidification of the polysulfide ions by sulfide to elemental S (Steudel, 1996). Elemental S molecules tend to form clusters with a hydrophobic nature and thus precipitate in the solid matrix. Reduction of Fe^{3+} ions to Fe^{2+} by sulfide and subsequent formation of FeS minerals also involves the production of elemental S (Shakeri Yekta et al., 2017). Furthermore, microbial oxidation of sulfide is a possible pathway of elemental S production in anaerobic environments. This process is usually coupled to nitrate reduction, in which nitrate serves as electron acceptor and sulfide as electron donor (Sher et al., 2008; Yang et al., 2005). Two major mechanisms are proposed for the formation of pyrite in Fe–S-water systems, so-called "H_2S" and "polysulfide" mechanisms. The "H_2S" mechanism involves a reaction of H_2S with FeS and formation of inner sphere $[Fe-S-SH_2]$ intermediate complexes (Rickard & Luther III, 1997). The "polysulfide" mechanism involves an addition of elemental S to FeS molecules with the formation of iron polysulfide intermediates (Luther III, 1991). Accordingly, formation of pyrite in anaerobic digesters is possible due to the presence of the pyrite precursors. However, nucleation and precipitation of pyrite are kinetically constrained (Rickard & Luther III, 1997) and the presence of zero-valent S in the anaerobic digesters likely relates to the formation of elemental S and polysulfides. Little is known about the effects of operational parameters such as mixing rate and hydraulic retention time on formation, crystal growth and aging of S-containing minerals in anaerobic digesters.

Reduced organic S fraction represents inherited S-containing organic compounds such as S-containing amino acids, which are found in almost all types of cells. Furthermore, sulfidic condition of the anaerobic digesters may promote in situ processes such as incorporation of S to organic molecules through sulfurization.

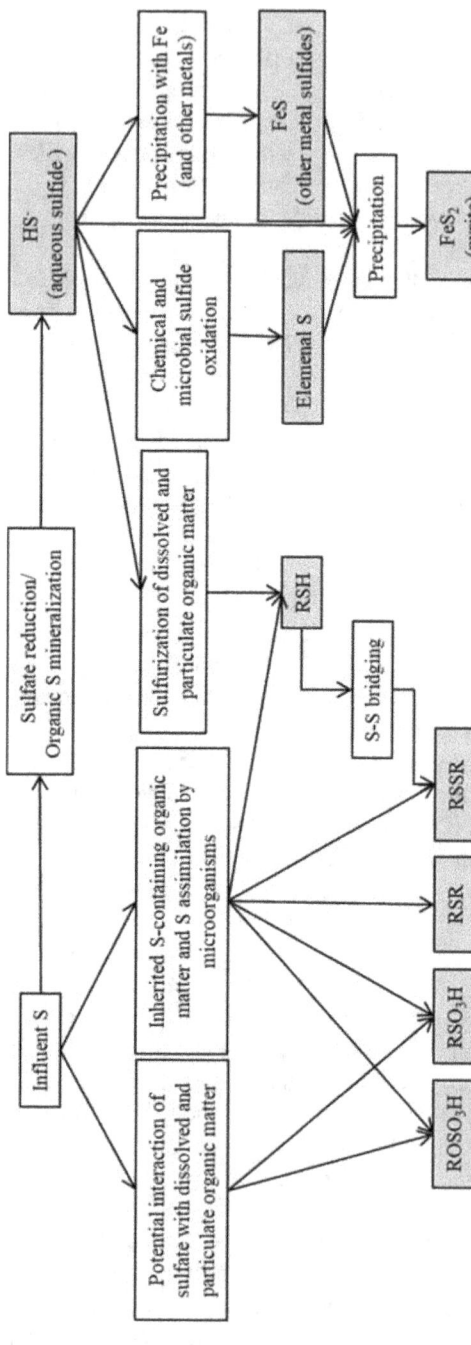

Figure 1.6 An overview of the S cycle in anaerobic digesters. Grey blocks represent various S species identified based on chemical speciation studies. White blocks represent potential precursors/processes, which contribute to occurrence of different S species in anaerobic digesters.

In this process, active sites of labile organic molecules are "quenched" by sulfide, resulting in formation of thiol groups (Eglinton *et al.*, 1994; Vairavamurthy & Mopper, 1987). The prerequisite for sulfurization of organic molecules is a low concentration of reactive Fe and the abundance of free sulfide ions. Characterization of dissolved organic matter in industrial co-digesters and sewage sludge digesters showed that a large number of organic molecules in the liquid phase contains S in their molecular structures (Shakeri Yekta *et al.*, 2012). The diversity of these molecules in terms of chemical properties suggested that the pool of dissolved organic S in anaerobic digesters might partially be a result of the organic matter sulfurization. Presence of disulfide species might be due to the availability of thiol active sites, which tend to form mono- and disulfidic S bridges (Sinninghe Damste & De Leeuw, 1990).

Figure 1.6 illustrates an overview of S cycle, which is conceptualized based on the major S species in anaerobic digesters. In summary, microbial sulfate reduction and degradation of S-containing organic matter results in a conversion of influent S content to sulfide, which is the main precursor of the reduced S species in anaerobic digesters. Sulfide reacts with Fe (and other metals) and precipitates mainly as FeS in solid phase. When sulfide reacts with potential toxic metals, the formation of metal sulfide precipitation may alleviate metal toxicity for the AD microbial community such as methanogens (Paulo *et al.*, 2015, 2017). Sulfide is also oxidized to zero-valent S, which likely represents elemental and polysulfidic S formation via chemical and (potentially) microbial processes. Pyrite formation may as well contribute to S cycle in anaerobic digesters, yet the kinetic limitations may constrain the formation of pyrite at large quantities. Inherited S-containing organic matter (e.g. S-containing amino acids in biomass and undegraded organic matter), S assimilation by microorganisms, and sulfurization of the organic matter contribute to the pool of dissolved and particulate reduced organic S (i.e. RSH, RSR, and RSSR). Minor groups of oxidized S including organic compounds with C–O–S and C–S linkages are also present, which may originate from influent organic sulfate species.

1.7 MAJOR ELEMENTS BIOGEOCHEMISTRY INTERPLAY IN ANAEROBIC DIGESTERS

Iron has been reported to play an important role in immobilizing phosphorus in soil and sediments through the formation of FePs. The mobilization and immobilization of phosphorus from FePs in these natural systems, in response to changes of ORP, is well documented. Similar processes have been reported to occur in WWTP (aerobic as well as anaerobic processes) (see Wilfert *et al.*, 2015 for a complete overview of iron and phosphorus cycle in WWTP). Indeed, WWTP processes operate in a large range of ORP to allow different microbial processes to take place. The ORP in a WWTP may range from less than -300 mV, during AD to more than $+200$ mV during the nitrification process. Consequently, microbial and chemical processes can take place and alter FePs stability by oxidizing or reducing iron or by replacing phosphorus with sulfide or other ions. Changes in the ORP in both, positive and negative ranges and subsequent changes in microbial processes can assist in either retaining or mobilizing phosphorus from FePs.

The chemical or biological reductive dissolution of ferric iron can cause iron-bound phosphorus (i.e. Fe(III)oxide–P complex and Fe(III)P mineral) to be released (Figure 1.7). Dissimilatory iron-reducing bacteria reduce ferric iron in iron oxides or iron phosphate minerals, thereby mobilizing phosphorus. Once formed, ferrous iron can precipitate as secondary iron oxides (e.g. magnetite or green rust) or as ferrous iron phosphate minerals (e.g. vivianite). Wilfert *et al.* (2018) reported that one can take advantage of the free paramagnetic nature of vivianite particles to recover them from digested sludge using magnetic separators (Wilfert *et al.*, 2018). In contrast, when electron acceptors are present (e.g. oxygen or nitrate), dissolved or solid ferrous iron compounds may be oxidized. Biogenic iron oxides that can be formed in the presence of iron-oxidizing bacteria include goethite, magnetite, ferrihydrite, and green rust. Biogenic iron oxides are often amorphous and nanocrystalline and display high orthophosphate binding capacities (Buliauskaité *et al.*, 2019). As mentioned in the Section 1.6, sulfide, biologically produced by sulfate-reducing bacteria and mineralization of organic S, can abiotically reduce ferric iron compounds and can further react to form various iron sulfide compounds and mobilize Fe-bound phosphorus. Also, sulfide has been reported to solubilize phosphorus selectively from FePs (Fe(II)/Fe(III) oxide–P complex and Fe(II)P mineral) containing sludge, leading to the formation

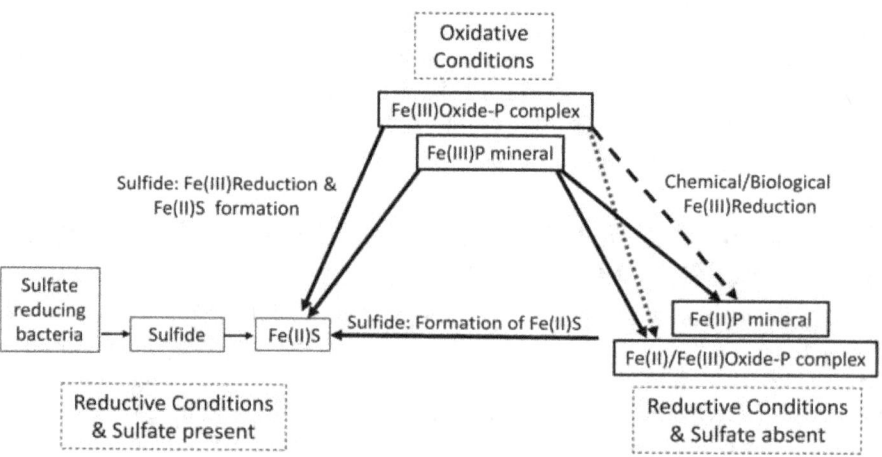

Figure 1.7 Redox processes and the cycling of P. The arrows represent the effect on soluble P: → implies P release, ---→ implies P sink, ·······→ implies not clear.

of ferrous sulfide as the expense of ferrous phosphate precipitates (vivianite) (Roussel & Carliell-Marquet, 2016; Takashima, 2018; Wilfert *et al.*, 2015). The addition of iron in an anaerobic digester may be seen as a way to prevent sulfide reacting with phosphorus precipitate and therefore favor phosphorus recovery as vivianite (Korving *et al.*, 2019; Wilfert *et al.*, 2016). Flores-Alsina *et al.* (2016) recently proposed a series of extensions to functionally upgrade the IWA Anaerobic Digestion Model No. 1 (ADM1) to allow for plant-wide phosphorus simulation. Due to the close interplay between the phosphorus, sulfur and iron cycles a substantial increase in model complexity due to the involved three-phase physico-chemical and biological transformations has been set (Batstone *et al.*, 2018). The same model has been implemented to assess the effect of zero valent iron (ZVI) addition during the AD of waste activated sludge from WWTP. Puyol *et al.* (2017) has confirmed by modeling experimental data that P-recovery potential is considerably reduced as soluble P decreased by one order of magnitude in the ZVI-amended digester compared with the non-amended digester and the phosphorus was mainly recovered in the solid digestate as vivianite.

1.8 CONCLUDING REMARKS

Depending on the composition of feedstock processed, AD may suffer from low methane yield, instability and even process failure due to high VFA and ammonia concentrations leading to pH and toxic shocks. This aspect is strongly connected with the carbon and nitrogen cycles and implies that feedstock composition, as well as the operational conditions of AD, should be carefully selected in order to ensure the sustainable production of methane. This aspect is further complicated by trace elements status in the feedstock especially when AD is operated on a mono-substrate. The geochemical cycle of sulfur, iron and phosphorus is currently the focus of numerous research activities aiming on one hand to develop new avenues for the recovery of phosphorus from sewage sludge and the improvement of AD performance with the addition of iron minerals in different chemical forms. As these three elements are strongly interacting with each other and their behavior is strongly affected by various redox reactions, a detailed understanding of their cycle and mineralogy in anaerobic digesters is needed. This statement is further reinforced if we consider that these major elements also affect the speciation and bioavailability of biologically active trace elements which are known to be of very high importance for AD performance.

ACKNOWLEDGEMENTS

The authors gratefully acknowledge EU COST Action ES1302 "Ecological functions of trace metals in anaerobic biotechnologies" for making this collaborative project possible.

REFERENCES

Akunna J. C., Bizeau C. and Moletta R. (1992). Denitrification in anaerobic digesters: possibilities and influence of wastewater COD/N-NOx ratio. *Environmental Technology*, **13**, 825–836.

Angelidaki I., Karakashev D., Batstone D. J., Plugge C. M. and Stams A. J. (2011). Biomethanation and its potential. *Methods in Enzymology*, **494**, 327–351.

Batstone D. J., Flores-Alsina X. and Hauduc H. (2018). Modeling the phosphorus cycle in the wastewater treatment process. In: Phosphorus: Polluter and Resource of the Future: Motivations, Technologies and Assessment of the Elimination and Recovery of Phosphorus from Wastewater, C. Schaum (ed.), IWA Publishing, London, UK, pp. 219–238.

Bayrakdar A., Sürmeli R. Ö. and Çalli B. (2017). Dry anaerobic digestion of chicken manure coupled with membrane separation of ammonia. *Bioresource Technology*, **244**, 816–823.

Bernhard A. (2010). The nitrogen cycle: processes, players, and human impact. *Nature Education Knowledge*, **3**(10), 25.

Buliauskaitė R., Wilfert P., Suresh Kumar P., de Vet W. W., Witkamp G. J., Korving L. and van Loosdrecht M. C. (2019). Biogenic iron oxides for phosphate removal. *Environmental Technology*, 1–7. DOI: 10.1080/09593330.2018.1496147.

Byrne B. and Goldblatt C. (2014). Radiative forcings for 28 potential Archean greenhouse gases. *Climate of the Past*, **10**, 1779–1801.

Calli B., Mertoglu B., Inanc B. and Yenigun O. (2005). Effects of high free ammonia concentrations on the performances of anaerobic bioreactors. *Process Biochemistry*, **40**(3–4), 1285–1292.

Chen S., Zhang J. and Wang X. (2015). Effects of alkalinity sources on the stability of anaerobic digestion from food waste. *Waste Management & Research*, **33**(11), 1033–1040.

Demirel B. and Scherer P. (2011). Trace element requirements of agricultural biogas digesters during biological conversion of renewable biomass to methane. *Biomass and Bioenergy*, **35**(3), 992–998.

Dong J., Zhao Y., Hong M. and Zhang W. H. (2009). Influence of alkalinity on the stabilization of municipal solid waste in anaerobic simulated bioreactor. *Journal of Hazardous Materials*, **163**, 717–722.

Eglinton T. I., Irvine J. E., Vairavamurthy A., Zhou W. and Manowitz B. (1994). Formation and diagenesis of macromolecular organic sulfur in Peru margin sediments. *Organic Geochemistry*, **22**, 781–799.

Feng Y., Zhang Y., Quan X. and Chen S. (2014). Enhanced anaerobic digestion of waste activated sludge digestion by the addition of zero valent iron. *Water Research*, **52**, 242–250.

Flores-Alsina X., Solon K., Mbamba C. K., Tait S., Gernaey K. V., Jeppsson U. and Batstone D. J. (2016). Modelling phosphorus (P), sulfur (S) and iron (Fe) interactions for dynamic simulations of anaerobic digestion processes. *Water Research*, **95**, 370–382.

Gallert C., Bauer S. and Winter J. (1998). Effect of ammonia on the anaerobic degradation of protein by a mesophilic and thermophilic biowaste population. *Applied Microbiology and Biotechnology*, **50**, 495–501.

Gonzalez-Estrella J., Puyol D., Sierra-Alvarez R. and Field J. A. (2015). Role of biogenic sulfide in attenuating zinc oxide and copper nanoparticle toxicity to acetoclastic methanogenesis. *Journal of Hazardous Materials*, **283**, 755–763.

Güngör K. and Karthikeyan K. G. (2005). Probable phosphorus solid phases and their stability in anaerobically digested dairy manure. *Transactions of the American Society of Agricultural Engineers*, **48**(4), 1509–1520.

Gustavsson J., Shakeri Yekta S., Sundberg C., Karlsson A., Ejlertsson J., Skyllberg U. and Svensson B. H. (2013). Bioavailability of cobalt and nickel during anaerobic digestion of sulfur-rich stillage for biogas formation. *Applied Energy*, **112**, 473–477.

Hansen K. H., Angelidaki I. and Ahring B. K. (1998). Anaerobic digestion of swine manure: inhibition by ammonia. *Water Research*, **32**(1), 5–12.

Hundal L. S., Carmo A. M., Bleam W. L. and Thompson M. L. (2000). Sulfur in biosolids-derived fulvic acid: characterization by XANES spectroscopy and selective dissolution approaches. *Environmental Science and Technology*, **34**, 5184–5188.

Ilbert M. and Bonnefoy V. (2013). Insight into the evolution of the iron oxidation pathways. *Biochimica et Biophysica Acta (BBA)-Bioenergetics*, **1827**(2), 161–175.

Jetten M. S., Niftrik L. V., Strous M., Kartal B., Keltjens J. T. and Op den Camp H. J. (2009). Biochemistry and molecular biology of anammox bacteria. *Critical Reviews in Biochemistry and Molecular Biology*, **44**(2–3), 65–84.

Jiang Y., Heaven S. and Banks C. J. (2012). Strategies for stable anaerobic digestion of vegetable waste. *Renewable Energy*, **44**, 206–214.

Kamp A., Høgslund S., Risgaard-Petersen N. and Stief P. (2015). Nitrate storage and dissimilatory nitrate reduction by eukaryotic microbes. *Frontiers in Microbiology*, **6**, 1492.

Kanders L., Beier M., Nogueira R. and Nehrenheim E. (2018). Sinks and sources of anammox bacteria in a wastewater treatment plant – screening with qPCR. *Water Science and Technology*, **78**, 441–451.

Kappler A., Emerson D., Gralnick J. A., Roden E. E. and Muehe E. M. (2015). Geomicrobiology of iron. *Ehrlich's Geomicrobiology*, 343–400.

Kayhanian M. (1994). Performance of a high-solids anaerobic digestion process under various ammonia concentrations. *Journal of Chemical Technology & Biotechnology*, **59**, 349–352.

Korving L., Van Loosdrecht M. and Wilfert P. (2019). Effect of iron on phosphate recovery from sewage sludge. In: Phosphorus Recovery and Recycling, Springer, Singapore, pp. 303–326.

Li B., Boiarkina I., Yu W., Huang H. M., Munir T., Wang G. Q. and Young B. R. (2019). Phosphorous recovery through struvite crystallization: challenges for future design. *Science of the Total Environment*, **648**, 1244–56.

Liu Y., Zhang Y. and Ni B. J. (2015). Zero valent iron simultaneously enhances methane production and sulfate reduction in anaerobic granular sludge reactors. *Water Research*, **75**, 292–300.

Lovley D. R. (1997). Microbial Fe(III) reduction in subsurface environments. *FEMS Microbiology Reviews*, **20**(3–4), 305–313.

Luther G. W., III (1991). Pyrite synthesis via polysulfide compounds. *Geochimica et Cosmochimica Acta*, **55**, 2839–2849.

Merchant S. S. and Helmann J. D. (2012). Elemental economy. Microbial strategies for optimizing growth in the face of nutrient limitation. *Advances in Microbial Physiology*, 91–210.

Möller K. and Müller T. (2012). Effects of anaerobic digestion on digestate nutrient availability and crop growth: a review. *Engineering in Life Sciences*, **12**(3), 242–257.

Mortimer R. J. G., Krom M. D., Harris S. J., Hayes P., Davies I. M., Davison W. and Zhang H. (2002). Evidence for suboxic nitrification in recent marine sediments. *Marine Ecology Progress Series*, **236**, 31–35.

Mulder A., Van de Graaf A. A., Robertson L. A. and Kuenen J. G. (1995). Anaerobic ammonium oxidation discovered in a denitrifying fluidized bed reactor. *FEMS Microbiology Ecology*, **16**(3), 177–183.

O'Flaherty V., Colohan S., Mulkerrins D. and Colleran E. (1999). Effect of sulphate addition on volatile fatty acid and ethanol degradation in an anaerobic hybrid reactor. II: Microbial interactions and toxic effects. *Bioresource Technology*, **68**, 109–120.

Paulo L. M., Stams A. J. and Sousa D. Z. (2015). Methanogens, sulphate and heavy metals: a complex system. *Reviews in Environmental Science and Bio/Technology*, **14**(4), 537–553.

Paulo L. M., Ramiro-Garcia J., van Mourik S., Stams A. J. and Sousa D. Z. (2017). Effect of nickel and cobalt on methanogenic enrichment cultures and role of biogenic sulfide in metal toxicity attenuation. *Frontiers in Microbiology*, **8**, 1341.

Puyol D., Flores-Alsina X., Segura Y., Molina R., Jerez S., Gernaey K. V. and Martinez F. (2017). ZVI addition in continuous anaerobic digestion systems dramatically decreases P recovery potential: dynamic modelling. Frontiers – International Conference on Wastewater Treatment and Modeling, pp. 211–217, Springer, Cham.

Raskin L., Rittmann B. and Stahl D. (1996). Competition and coexistence of sulfate-reducing and methanogenic populations in anaerobic biofilms. *Applied and Environmental Microbiology*, **62**, 3847–3857.

Rickard D. and Luther G. W., III (1997). Kinetics of pyrite formation by the H2S oxidation of iron(II) monosulfide in aqueous solutions between 25°C and 125°C: the mechanism. *Geochimica et Cosmochimica Acta*, **61**, 135–147.

Romero-Güiza M. S., Vila J., Mata-Alvarez J., Chimenos J. M. and Astals S. (2016). The role of additives on anaerobic digestion: a review. *Renewable and Sustainable Energy Reviews*, **58**, 1486–1499.

Roussel J. and Carliell-Marquet C. (2016). Significance of vivianite precipitation on the mobility of iron in anaerobically digested sludge. *Frontiers in Environmental Science*, **4**, 60.

Shakeri Yekta S., Gonsior M., Schmitt-Kopplin P. and Svensson B. H. (2012). Characterization of dissolved organic matter in full scale continuous stirred tank biogas reactors using ultrahigh resolution mass spectrometry: a qualitative overview. *Environmental Science and Technology*, **46**, 12711–12719.

Shakeri Yekta S., Svensson B. H., Björn A. and Skyllberg U. (2014). Thermodynamic modeling of iron and trace metal solubility and speciation under sulfidic and ferruginous conditions in full scale continuous stirred tank biogas reactors. *Applied Geochemistry*, **47**, 61–73.

Shakeri Yekta S., Ziels R. M., Björn A., Skyllberg U., Ejlertsson J., Karlsson A., Svedlund M., Willén M. and Svensson B. H. (2017). Importance of sulfide interaction with iron as regulator of the microbial community in biogas reactors and its effect on methanogenesis, volatile fatty acids turnover, and syntrophic long-chain fatty acids degradation. *Journal of Bioscience and Bioengineering*, **123**, 597–605.

Sher Y., Schneider K., Schwermer C. U. and van Rijn J. (2008). Sulfide-induced nitrate reduction in the sludge of an anaerobic digester of a zero-discharge recirculating mariculture system. *Water Research*, **42**, 4386–4392.

Sinninghe Damste J. S. and De Leeuw J. W. (1990). Analysis, structure and geochemical significance of organically-bound sulphur in the geosphere: state of the art and future research. *Organic Geochemistry*, **16**, 1077–1101.

Sommers L. E., Tabatabai M. A. and Nelson D. W. (1977). Forms of sulfur in sewage sludge. *Journal of Environmental Quality*, **6**, 42–46.

Speece R. E. and McCarty P. L. (1964). Nutrient requirements and biological solids accumulation in anaerobic digestion. *Advances in Water Pollution Research*, **2**, 305–322.

Steudel R. (1996). Mechanism for the formation of elemental sulfur from aqueous sulfide in chemical and microbiological desulfurization processes. *Industrial and Engineering Chemistry Research*, **35**, 1417–1423.

Strohm T. O., Griffin B., Zumft W. G. and Schink B. (2007). Growth yields in bacterial denitrification and nitrate ammonification. *Applied and Environmental Microbiology*, **73**(5), 1420–1424.

Stüeken E. E., Kipp M. A., Koehler M. C. and Buick R. (2016). The evolution of Earth's biogeochemical nitrogen cycle. *Earth-Science Reviews*, **160**, 220–239.

Sun C., Cao W., Banks C. J., Heaven S. and Liu R. (2016a). Biogas production from undiluted chicken manure and maize silage: a study of ammonia inhibition in high solids anaerobic digestion. *Bioresource Technology*, **218**, 1215–1223.

Sun H., Wu S. and Dong R. (2016b). Monitoring volatile fatty acids and carbonate alkalinity in anaerobic digestion: titration methodologies. *Chemical Engineering & Technology*, **39**(4), 599–610.

Takashima M. (2018). Enhanced Phosphate Release from Anaerobically Digested Sludge Through Sulfate Reduction. *Waste and Biomass Valorization*, 1–7.

Takashima M. and Speece R. E. (1989). Mineral nutrient requirements for high-rate methane fermentation of acetate at low SRT. *Research Journal of the Water Pollution Control Federation*, **61**(11–12), 1645–1650.

Vairavamurthy A. and Mopper K. (1987). Geochemical formation of organosulphur compounds (thiols) by addition of H2S to sedimentary organic matter. *Nature*, **329**, 623–625.

Van de Graaf A. A., Mulder A., de Bruijn P., Jetten M. S., Robertson L. A. and Kuenen J. G. (1995). Anaerobic oxidation of ammonium is a biologically mediated process. *Applied and Environmental Microbiology*, **61**(4), 1246–1251.

van Hullebusch E., Rossano S., Farges F., Lenz M., Labanowski J., Lagarde P., Flank A.M. and Lens P. (2009). Sulfur K-edge XANES spectroscopy as a tool for understanding sulfur chemical state in anaerobic granular sludge. *Journal of Physics: Conference Series*, **190**.

van Lier J. B., Mahmoud N. and Zeeman G. (2008). Anaerobic wastewater treatment. Biological wastewater treatment, principles, modelling and design. In: Biological Wastewater Treatment, M. Henze, M. C. van Loosdrecht, G. A. Ekama and D. Brdjanovic (eds), IWA Publishing, London, UK, 415–456.

Van Rensburg P., Musvoto E. V., Wentzel M. C. and Ekama G. A. (2003). Modelling multiple mineral precipitation in anaerobic digester liquor. *Water Research*, **37**(13), 3087–3097.

Ward B. (2012). The global nitrogen cycle. *Fundamentals of Geobiology*, **1**, 36–48.

Wei J., Hao X., van Loosdrecht M. C. and Li J. (2018). Feasibility analysis of anaerobic digestion of excess sludge enhanced by iron: a review. *Renewable and Sustainable Energy Reviews*, **89**, 16–26.

Wei Q., Zhang W., Guo J. and Wu S. (2014). Performance and kinetic evaluation of a semi-continuously fed anaerobic digester treating food waste: effect of trace elements on the digester recovery and stability. *Chemosphere*, **117**, 477–485.

Wellinger A., Murphy J. D. and Baxter D. (eds) (2013). The Biogas Handbook: Science, Production and Applications. Elsevier, the Netherlands.

Wilfert P., Kumar P. S., Korving L., Witkamp G. J. and van Loosdrecht M. C. (2015). The relevance of phosphorus and iron chemistry to the recovery of phosphorus from wastewater: a review. *Environmental Science & Technology*, **49**(16), 9400–9414.

Wilfert P., Mandalidis A., Dugulan A. I., Goubitz K., Korving L., Temmink H. and Van Loosdrecht M. C. M. (2016). Vivianite as an important iron phosphate precipitate in sewage treatment plants. *Water Research*, **104**, 449–460.

Wilfert P., Dugulan A. I., Goubitz K., Korving L., Witkamp G. J. and Van Loosdrecht M. C. M. (2018). Vivianite as the main phosphate mineral in digested sewage sludge and its role for phosphate recovery. *Water Research*, **144**, 312–321.

Yang W., Vollertsen J. and Hvitved-Jacobsen T. (2005). Anoxic sulfide oxidation in wastewater of sewer networks. *Water Science and Technology*, **1**, 191–199.

Yang Y., Zhang C. and Hu Z. (2013). Impact of metallic and metal oxide nanoparticles on wastewater treatment and anaerobic digestion. *Environmental Science: Processes & Impacts*, **15**(1), 39–48.

Zhang L., Lee Y. W. and Jhang D. (2011). Anaerobic co-digestion of food waste and piggery wastewater: focusing on the role of trace elements. *Bioresource Technology*, **102**, 5048–5059.

2

Anaerobic Digesters, Trace Elements and Biogeochemical Processes

Lucian C. Staicu[1], Stephane Simon[2], Gilles Guibaud[2], Sepehr Shakeri Yekta[3], Baris Calli[4], Jan Bartacek[5], Fernando G. Fermoso[6] and Eric D. van Hullebusch[7,8]

[1]*Faculty of Biology, University of Warsaw, Warsaw, Poland*
[2]*Faculté des Sciences et Techniques, Université de Limoges, Pereine Ura Irstea, Limoges, France*
[3]*Department of Thematic Studies-Environmental Change, Linköping University, Linköping, Sweden*
[4]*Environmental Engineering Department, Marmara University, Istanbul, Turkey*
[5]*Department of Water Technology and Environmental Engineering, Institute of Chemical Technology Prague, Prague, Czech Republic*
[6]*Instituto de la Grasa (C.S.I.C.), Sevilla, Spain*
[7]*Department of Environmental Engineering and Water Technology, IHE Delft Institute for Water Education, Delft, The Netherlands*
[8]*Institut de Physique du Globe de Paris, Sorbonne Paris Cité, Université Paris Diderot, UMR 7154, CNRS, F-75005 Paris, France*

ABSTRACT

Essential trace elements (TE) are a prerequisite that ensures optimal performance of the anaerobic digestion (AD) process. However, finding the proper way to deliver these micronutrients to microbial communities is not an easy task. The chemical speciation of TE and the complex environment characterizing AD play a critical role in their mobility, bioavailability, and toxicity. These aspects are particularly critical when establishing the total versus bioavailable concentration of TE, by properly balancing the two sides of the same coin, namely essentiality and toxicity. Both non-redox sensitive (e.g. Co, Cu, Ni, Zn) and redox-sensitive (e.g. Fe, Mn, Mo, Se, W) elements engage in a complex interplay with the mineral and organic phases present in AD. In addition, TE can also interact with each other, thus further complicating our current understanding. All these 'parasitic' reactions may render a large fraction of supplemented TE non-bioavailable for the efficient degradation of organic matter by microbial consortia, therefore limiting the biomethane yield. Current analytical limitations related to sampling and assessing the speciation, bioavailability, and matrix (liquid/solid) analysis add to the difficulty of understanding the bigger picture. This chapter reviews and discusses at length all these aspects, providing an up-to-date presentation of the biogeochemistry of TE in AD.

KEYWORDS: anaerobic digestion, biogas, biogeochemistry, resource recovery, trace elements

2.1 INTRODUCTION

Anaerobic digestion (AD) is a biological process commonly used for the treatment of organic waste streams and production of methane as a renewable energy carrier. The effluents from anaerobic digesters, so called digestate, contain high levels of nutrients such as nitrogen (N) and phosphorous (P) and can be applied as a sustainable alternative to conventional mineral fertilizers in agricultural practices (Monlau *et al.*, 2015). Efficient AD of organic substrates requires a balanced amount of trace elements (TE) for growth and activities of the diverse microbial consortia involved in AD of organic matter (Schnürer, 2016). Since the feedstock of AD processes may lack sufficient amounts of TE, a constant TE supplementation by operators is often required to assure the stability and efficiency of the AD processes (Demirel & Scherer, 2011). It has been well documented that the absence of key TE results in a serious underperformance of the AD processes (e.g. Banks *et al.*, 2012; Gustavsson *et al.*, 2011; Jarvis *et al.*, 1997; Moestedt *et al.*, 2015; Molaey *et al.*, 2018a; Schmidt *et al.*, 2014). Several TE such as iron (Fe), nickel (Ni), cobalt (Co), molybdenum (Mo), selenium (Se), tungsten (W), or zinc (Zn) are identified as essential additives to the AD processes in which TE deficient feedstock is treated (Glass & Orphan, 2012).

A number of TE, including Fe, Mn, Mo, Se, or W, may exist under several valence states, implying that redox conditions play an important role in the regulation of their chemical speciation. Furthermore, non-redox sensitive TE, including Co, Ni, or Zn, often demonstrate a restricted mobility under anaerobic conditions as a result of their limited valence-state interchange (Shakeri Yekta *et al.*, 2014). Once introduced in the anaerobic system, redox-sensitive and non-redox sensitive TE are involved in a complex network of reactions, such as oxidation-reduction, precipitation, co-precipitation, adsorption, and ion exchange, which may limit their availability for microorganisms (Fermoso *et al.*, 2015). Accordingly, TE concentrations supplemented to AD processes often exceed the 'trace levels' required by microorganisms (Pinto-Ibieta *et al.*, 2016). This raises the challenging question of TE addition optimization, which requires an understanding of the biochemical processes regulating the chemical speciation and bioavailability of TE in AD processes. Over-supplementation of TE is counterproductive as it: (i) entails additional process costs; (ii) elicits toxicity of TE with low toxicity threshold concentrations (e.g. Se); and (iii) poses restriction for down-stream application of digestate as fertilizer/soil conditioner due to its TE contents.

The purpose of this chapter is to review and discuss the current state of knowledge on the biogeochemistry of redox sensitive and non-redox sensitive TE with relevance in AD. In addition, important aspects related to total versus bioavailable TE concentration and the challenge to assess speciation and bioavailability of TE for optimum dosing to anaerobic digesters will be discussed.

2.2 TOTAL VERSUS BIOAVAILABLE TRACE ELEMENTS

The determination of total TE concentration is important when one is seeking to evaluate the potential effect of the deficiency or excess of TE on AD processes. However, it has been reported that the total TE concentration is a poor indicator of the elemental fraction available to microorganisms (Thanh *et al.*, 2016). As shown in Table 2.1, the optimal TE concentrations required for optimal operating conditions in anaerobic bioreactors, based on total TE content, differ by as much as four orders of magnitude (Schattauer *et al.*, 2011). However, the chemical form of the reported elements is not considered, which obviously has an important impact on TE bioavailability as discussed by Thanh *et al.* (2016).

In order to illustrate the importance of TE speciation on methanogens activity, Bartacek *et al.* (2008) investigated the influence of Co speciation on its toxicity to methylotrophic methanogenesis in anaerobic granular sludge. The Co speciation in three different media that contained varying concentrations of complexing ligands (carbonates (CO_3^{2-}), phosphates (PO_4^{3-}) and ethylenediaminetetraacetic acid (EDTA)) was studied. Three Co fractions (total Co added, dissolved Co and free Co ion) were measured in the liquid media and were correlated with data from batch toxicity experiments. The average concentration of Co that was required for 50% inhibition of methanogenic activity (IC50) for free Co^{2+} in the three sets of measurements was about 1.3×10^{-5} mol L^{-1} in the three different media. Complexation (and/or precipitation) with EDTA, PO_4^{3-} and CO_3^{2-} was shown to decrease the toxicity of Co on methylotrophic methanogenesis. However, the free Co concentration is considered as the key parameter to correlate with Co toxicity onto methanogens, as this metal species is defined as being fully bioavailable. Such a conclusion

Table 2.1 TE concentration measured or calculated for optimal operating conditions in AD (modified from Schattauer et al. 2011).

µM (Fresh Matter)

	Sahm (1981)[a]	Takashima et al. (1990)[b]	Kloss (1986)[c]	Weiland (2006)[d]	Seyfried et al. (1990)[e]	Mudrack & Kunst (2003)[f]	Pobeheim et al. (2010)[g]	Bischofsberger et al. (2005)[h]	Schattauer et al. (2011)[i]
B	$0.092–1.02 \times 10^3$								79.5–442.1
Ca		>13.5–998.1							
Co	>0.01–2.0	1.0	8.5–339	0.05–1.02	0.05–1.0	0.05–1.0	0.4–169.7	1.02	0.5–27.8
Cu	0.9–1007.2								0.4–25.8
Fe	>5–902.6		$179–3.6 \times 10^3$	17.9–179	17.9–179	17.9–170			$859–1.09 \times 10^5$
Mg	$14.8 \times 10^3–1.97 \times 10^5$								$2.7 \times 10^3–1.65 \times 10^5$
Mn	0.09–1001							0.09–910	103.2–1354
Mo	>0.01–0.5	0.52	1.04–3.64	0.05–0.52	0.05–0.52	0.05–0.52	1.7–521	0.52	1.4–4.8
Ni	0.1–85.1	0.1	8.5–511	0.085–8.5	0.085–8.5	0.085–8.5	0.4–10.6	0.1	3.9–61.2
S	$9.98–5.1 \times 10^5$								$1.1 \times 10^3–7.2 \times 10^3$
Se(IV)	1–10		1.3–4.4		0.1			0.1	0.13–5.7
W	0.09–99.5		0.5–1.9		0.5–2.2				
Zn									131–1044

[a]Sahm (1981).
[b]Takashima et al. (1990).
[c]Kloss (1986).
[d]Weiland (2006).
[e]Seyfried et al. (1990).
[f]Mudrack & Kunst (2003).
[g]Pobeheim et al. (2010).
[h]Bischofsberger et al. (2005).
[i]Schattauer et al. (2011).

could be extrapolated to cases where concentrations of Co are added to stimulate the methylotrophic methanogens with a free Co ion concentration being optimal for a concentration in the range of 10^{-12}–10^{-13} mol L^{-1} (van Hullebusch *et al.*, personal communication).

The bioavailability is the degree to which elements are available for interaction with biological systems (Marcato *et al.*, 2009). These bioavailable TE can be divided into two categories by their uptake mechanisms: those with (i) active uptake i.e. internalization processes requiring direct metabolic activity from microorganisms to transfer TE through the plasma membrane; and (ii) passive uptake, i.e. uptake based only on a concentration gradient across the cell membrane. In addition to these processes, TE bioavailability is controlled by TE partitioning between the liquid and solid phases, and the diffusion of TE towards the microbial membrane surface. Once internalized, TE can impact the methane production yield via the intracellular, bioavailable fraction (Figure 2.1).

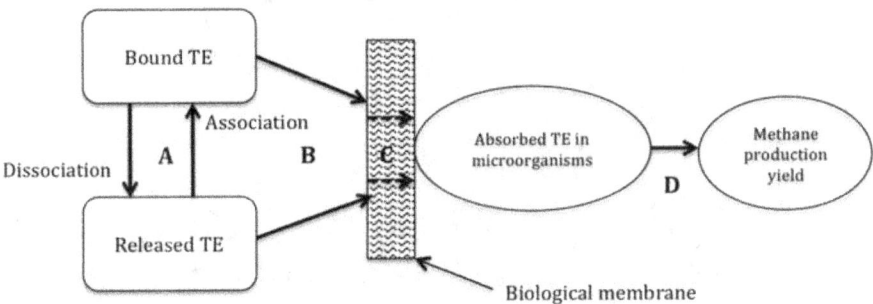

Figure 2.1 Simplified, conceptual representation of TE bioavailability in anaerobic digesters (adapted from NRC (2003) and reprinted from van Hullebusch *et al.*, 2016). **A**, **B** and **C** are related to bioavailability processes: TE interactions between phases, transport of TE to microorganisms and bio-uptake of TE through the biological membrane, respectively. **D** represents the biological response (i.e. methane production yield) as a function of the bioavailable TE intracellular concentration.

2.3 BIOGEOCHEMICAL PROCESSES
2.3.1 Introduction

In heterogeneous biological systems, such as engineered AD ecosystems, the distribution and fate of TE are controlled by a complex network of physical, chemical, and biological reactions (Fermoso *et al.*, 2009). The effect of these processes is a dynamic TE partitioning among different fractions: free TE ions, soluble organic and inorganic TE complexes, and TE bound to colloidal and biotic (microorganisms) particulate materials. Figure 2.2 shows the chemical reactions occurring both in liquid (i.e. TE reduction, precipitation or complexation) and solid phases (i.e. TE sorption in sludge) that play key roles in the chemical speciation of TE in AD bioreactors in relation to sulfur (S). In addition, the precipitation of metals by sulfide (S^{2-}), CO_3^{2-} and PO_4^{3-} may also play a pivotal role in nutrients and TE turnover (Figure 2.2) (Fermoso *et al.*, 2015; Maharaj *et al.*, 2018; Thanh *et al.*, 2016).

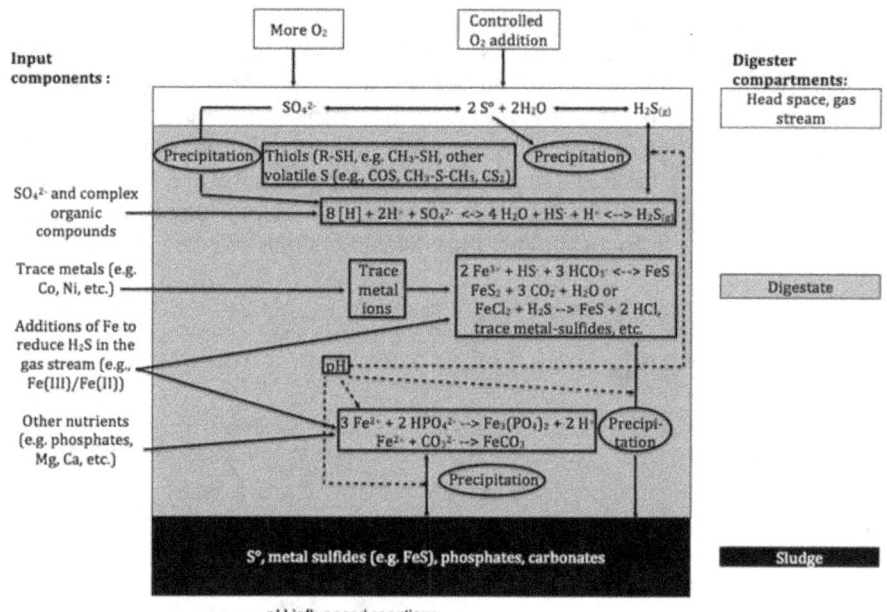

Figure 2.2 Sulfur turnover in biogas bioreactors and its influence on TE and nutrients speciation (modified from Möller and Müller (2012) and reprint from van Hullebusch *et al.*, 2016).

2.3.2 Non-redox sensitive elements

Non-redox sensitive metals such as Ni, Co, Zn or copper (Cu) occur in natural and engineered environments such as sediments or bioreactors (in vast majority) only in one valence state, most often +II. Even though other valences are known (e.g. Co(III) or Ni(I)), these forms are rare and their transformations (i.e. oxidation and reduction) do not occur under conditions prevailing in most natural environments.

Indirectly, redox potential influences the chemistry of non-redox sensitive TE by the occurrence of various redox sensitive anions. Specifically, S^{2-} dominates the chemistry of non-redox sensitive TE in anaerobic environments as it forms extremely strong precipitates. Under aerobic conditions, sulfide occurs only in trace amounts and sulfate does not form precipitates with TE.

All most important non-redox sensitive TE, i.e. Co, Ni, Zn and Cu, can form a great number of organic and inorganic complexes ranging from simple $[NiCl]^+$ to extremely complex cobalamins. This increases the bioavailability of non-redox sensitive TE in case of TE limitation, because complexes are much more mobile than precipitates. Even though the equilibrium concentrations of free ions are very low in the presence of these complexes, strong complexes such as EDTA have been shown to alleviate TE limitation (Fermoso et al., 2008).

Immobilization (or biosorption) of non-redox sensitive elements by anaerobic microorganisms, usually clustered in biofilms, includes a number of processes, most important of which are adsorption, ion exchange, absorption in extracellular polymers (EPS), precipitation and internalization (van Hullebusch et al., 2003). In real systems, it is difficult to distinguish individual processes and therefore the summary term biosorption is often used. Out of all processes included in the biosorption term, precipitation, absorption in EPS and internalization are probably the most significant.

An accurate description of TE precipitation in anaerobic biofilms is extremely challenging, and an experimental description of actual distribution is almost impossible. In most environments, S^{2-}, CO_3^{2-} and PO_4^{3-} anions are the most important precipitation agents for non-redox sensitive TE. In particular, S^{2-} complexes are extremely poorly soluble. On the contrary, CO_3^{2-} complexes are relatively soluble and therefore more bioavailable. TE are not completely biologically inert when precipitated with S^{2-}. Gustavsson et al. (2013a) observed an increase in methanogenic activity as the result of CoS dosing. Thus, S^{2-} precipitates with TE can, in some cases, create a pool of micronutrients which will be very slowly washed out from anaerobic bioreactors or sediments.

EPS play a crucial role in biosorption because they mostly contain anionic groups which form complexes with TE. The retention of TE in these complexes can then cause concentration gradient formation and, thus, regulate the exposure of bacteria to TE. TE bound in EPS can also play the role of a stock in the event of micronutrient deficiency in the environment (Fermoso et al., 2009). The strength of the bond between TE and the EPS depends on the conditions prevailing in the environment (e.g. pH, temperature) and the concentration of other TE present in the environment. At the same time, the properties of the given TE (e.g. electrical charge, radius) are very important.

2.3.2.1 Cobalt

Free cobalt ion (Co^{2+}) is present in natural and engineered environments in extremely low concentrations, i.e. at the level of pmol L^{-1} (Yekta et al., 2017). The vast majority of Co is present as inorganic complexes (e.g. $[Co(OH)]^+$ to $[Co(OH)_4]^{2-}$, $[CoCl]^+$, $[CoCO_3]^0$, $[CoHS]^+$ or $[CoSO_4]^0$), organic complexes (e.g. vitamin B_{12} or complexes with humic acids) and precipitates (mostly CoS (s), $CoCO_3$(s) and $Co(OH)_2$; Pitter, 2009).

Specifically in anaerobic environments, S^{2-} concentration controls the speciation of Co (Gustavsson *et al.*, 2013b) and only very strong complexing agents such as EDTA can keep Co in solution (Bartacek *et al.*, 2008).

As Co is an extremely important TE, especially for methanogenic organisms (Fermoso *et al.*, 2009), both bacteria and archaea (Marie Sych *et al.*, 2016; Rempel *et al.*, 2018) developed a way of Co extraction even from the environments where free Co ions are extremely rare, i.e. complexation of cobalt in cobalamins (or vitamin B_{12}). Besides their vital role in methyl transfer and radical generation (Matthews, 2001), they have been reported to play a crucial role in an active transport of cobalt across cell membranes (Rempel *et al.*, 2018; Zhang & Gladyshev, 2009). It has been shown by Fermoso *et al.* (2010) that supplementation of Co in the form of vitamin B_{12} is much more efficient than in any other form ([CoH$_2$EDTA], CoCl$_2$ etc.). It should be noted as well that improving bioavailability of Co might also lead to Co toxic concentrations (Pinto-Ibieta *et al.*, 2016).

2.3.2.2 Nickel

Nickel, similarly to cobalt, can form many different complexes such as $[Ni(OH)]^+$ to $[Ni(OH)_4]^{2-}$, $[NiCl]^+$, $[NiCO_3]^0$, $[NiHS]^+$ or $[NiSO_4]^0$. In anaerobic digesters, ammonia complexes are also important (from $[Ni(NH_3)]^{2+}$ to $[Ni(NH_3)_6]^{2+}$). Precipitates formed in most anaerobic environments are similar to Co precipitates, i.e. mainly NiS(s), NiCO$_3$(s) and Ni(OH)$_2$ (Pitter, 2009), with NiS(s) usually being vastly dominating.

2.3.2.3 Zinc

Zinc is also, most often, present in +II valence and its physical-chemical speciation is similar to Co and Ni. It forms similar complexes (with CO_3^{2-}, ammonia or hydroxyl ligands), but its complexing abilities are much weaker than for Co, Ni and Cu (Pitter, 2009). It also forms precipitates, again mainly with S^{2-}, CO_3^{2-} and hydroxyl ions (Legros *et al.*, 2017; Le Bars *et al.*, 2018).

2.3.2.4 Copper

Copper prevails in anaerobic environments as Cu(II), but it can be also present in small amounts as Cu(I). Its complexation ability is the strongest of all non-redox sensitive TE (Pitter, 2009). Again, Cu forms strong hydroxo-complexes and complexes with CO_3^{2-}, PO_4^{3-} and S^{2-}. It also forms extremely strong organic complexes. Similar to other non-redox sensitive TE, Cu precipitates with S^{2-}, CO_3^{2-}, and hydroxyl ions (Legros *et al.*, 2017).

2.3.3 Redox-sensitive elements

In contrast to non-redox sensitive elements, those displaying several valence states (i.e. redox sensitive) have higher mobility and display complex behavior with changing environmental conditions. Table 2.2 below compiles several redox-sensitive elements used in AD (Fe, Mn, Mo, Se, and W), together with their valence states and commonly reported form (compound). Of the five elements mentioned, Fe and Mn are commonly present in aquatic solutions as positively charged metal cations, whereas Mo, Se, and W form negatively charged oxyanions. When present in the same system, these elements often interact with each other in a complex interplay.

Table 2.2 Redox-sensitive elements commonly used in anaerobic digestion.

Chemical Element	Valence States*	AD Supplemented Valence State**	Form
Fe	+VI, +IV, +III, +II	+II	$FeCl_2$
Mn	+VII, +VI, +IV, +III, +II, +I	+II	$MnCl_2$
Mo	+VI, +III, +II	+VI	MoO_4^{2-} ***
Se	+VI, +IV, 0, −II	+VI, +IV	SeO_4^{2-}, SeO_3^{2-}
W	+VI, +V, +IV, +III, +II	+VI	WO_4^{2-}

*From Lide (2004); **Commonly reported in literature; ***Many articles report the addition of molybdenum as ammonium heptamolybdate: $(NH_4)_6Mo_7O_{24} \cdot 4H_2O$.

2.3.3.1 Iron

Iron is the fourth most abundant element on Earth and is usually found combined with oxygen as iron oxide minerals: hematite (Fe_2O_3), magnetite (Fe_3O_4), and siderite ($FeCO_3$) (Lide, 2004). Iron forms compounds of relevance in biology mainly in the +II (ferrous) and +III (ferric) oxidation states. The mineralogy of iron is complicated by its propensity to form non-stoichiometric compounds whose composition may vary. Complicating matters further, many coordination compounds of iron are known. Iron sulfide (FeS_2, pyrite) is an important compound in AD systems, especially in sulfidogenic environments where SO_4^{2-} is reduced to highly reactive S^{2-}, by sulfate-reducing bacteria (SRB). Sulfide is particularly important since it binds readily with Fe and other metallic cations yielding insoluble precipitates, thus rendering essential micronutrients biologically unavailable. Both FeO and FeS minerals play a major role in adsorption and precipitation phenomena involving other metals and metalloids present in AD. Iron is regularly added to AD systems as Fe(II) (Table 2.2), although the use of zero valent iron (ZVI) has also been documented (Zhang et al., 2011). ZVI may also act as an additional electron donor and buffer for the undissociated H_2S. Iron is involved in numerous biological processes (e.g. oxidation, transport, respiration) being an essential TE. It is noteworthy that iron-binding proteins are found in all living organisms.

In spite of its environmental abundance, Fe is a difficult element to exploit by living organisms as a consequence of its presence in insoluble states. Due to its importance for microbial metabolism, bacteria have evolved high-affinity sequestering molecules (siderophores) involved in extracellular Fe uptake (Neilands, 1995). Interestingly, certain siderophores are broad-range metal chelators (Braud et al., 2009). For instance, azotobactin, a fluorescent pyoverdine-like siderophore produced by *Azotobacter vinelandii*, binds molybdenum and vanadium (V) (Wichard et al., 2009).

2.3.3.2 Manganese

Manganese (Mn) minerals are widely distributed in nature: oxides, silicates, and carbonates being the most common. Pyrolusite (MnO_2) and rhodochrosite ($MnCO_3$) are commonly encountered Mn minerals (Lide, 2004). Manganese reacts with S^{2-}, forming insoluble MnS. As in the case of the other redox sensitive elements described in this chapter, the formation of MnS renders the micronutrients biologically unavailable. Interestingly, the addition of Mn oxides to AD systems can help mitigate ammonia presence. Ammonia is known to be an undesirable compound for methanogenic microbial consortia (Romero-Guiza et al., 2016). Various classes of enzymes have Mn cofactors, thus acting as an element of biological importance.

2.3.3.3 Molybdenum

Molybdenum (Mo) is a transition metal found in various valence states in minerals such as molybdenite (MoS_2), wulfenite ($PbMoO_4$), and powellite [$Ca(MoW)O_4$]. Industrially, Mo is a by-product of Cu and tungsten (W) smelting operations (Lide, 2004). Most molybdenum compounds have low water solubility, with the exception of molybdate oxyanion, MoO_4^{2-}.

Xu et al. (2005) investigated the adsorption of two major Mo species, molybdate (MoO_4^{2-}) and tetrathiomolybdate (MoS_4^{2-}), onto iron minerals pyrite (FeS_2) and goethite (FeOOH) and found the maximum sorption capacity in the acidic pH range (pH < 5). In addition, they observed a powerful competitive effect of phosphate for the adsorption sites of Fe minerals, while silicate and sulfate exhibited only a minor influence. MoS_4^{2-} showed the greatest propensity for Fe minerals, thus suggesting a possible mechanism of Mo immobilization under sulfidogenic and iron-rich conditions. In addition, MoS_4^{2-} was shown to strongly interact with organically modified montmorillonite (organo-smectite) at low pH (<5), displaying a maximum sorption capacity of 705 mmol kg^{-1} (Muir et al., 2017). The main sorption mechanisms proposed are chemisorption and ion exchange.

2.3.3.4 Selenium

Selenium (Se) occurs rarely as a pure element, being commonly associated with metal-sulfide minerals (e.g. pyrite and chalcopyrite) and biolites (e.g. coal, crude oil, bituminous shales). As such, Se is typically found together with Cu, Zn, lead (Pb), gold (Au), and Ni ores (Kyle $et\ al.$, 2011). Apart from the abiotic release of Se through rock weathering, volcanism and wildfires, the biotic component plays a major role in the mobilization of this element. Bacteria, possessing a versatile metabolism, have been documented to participate in almost all known valence-state transformations of Se (Staicu & Barton, 2017). In aquatic solutions, Se forms high valence state oxyanions: selenate, Se(VI), SeO_4^{2-}, and selenite, Se(IV), SeO_3^{2-}. These two Se oxyanions have a high toxicological potential in aquatic ecosystems, but under anaerobic conditions various bacteria can respire them to generate cellular energy (Staicu $et\ al.$, 2017). Unlike S, a closely related element, Se can be biologically reduced to elemental Se, Se^0, directly, a solid product, which is biologically unavailable (Figure 2.3a and Figure 2.4). It is noteworthy that Se can be incorporated into selenocysteine (Sec), the 21st proteinogenic amino acid, from its most reduced valence state, selenide, Se(-II). For a fuller discussion on the biosynthesis of Se, the reader is referred to Staicu $et\ al.$ (2017). Selenocysteine exists naturally in all three domains of life and, in selenoproteins, serves oxidoreductase functions against reactive oxygen species (Labunskyy $et\ al.$, 2014). Formate dehydrogenase (FDH), which plays an important role in syntrophic acetate oxidation in AD, is dependent on Se as well as Mo and W (Molaey $et\ al.$, 2018b). When Se is not supplemented to AD in its selenide form (Figure 2.3b), there is a significant risk of losing part of it as the biologically unavailable Se^0. Furthermore, biogenic Se^0 exhibits colloidal properties that make it environmentally persistent and prone to long distance transport in aquatic ecosystems when escaping the bioreactor setting, leading to secondary pollution (Cordoba & Staicu, 2018).

Selenium was shown to be complexed to organic matter through the formation of Se-metal–humic ternary complexes, although selenate appears to be considerably less reactive than selenite (Bruggeman $et\ al.$, 2007). On the other hand, at the mineral/water interface different processes involving Se can be taken into account such as adsorption, co-precipitation and surface precipitation processes (Fernandez-Martinez & Charlet, 2009). Adsorption of Se species at the mineral/water interface has been described to occur through both outer-sphere and inner-sphere complexation mechanisms. Selenium association with Fe, aluminum (Al) and Mn oxides and hydroxides has been reported in numerous articles (reviewed by Fernandez-Martinez & Charlet, 2009). In general, selenite is more reactive with various mineral surfaces and a possible explanation may be the inner-sphere complexation characterizing this interaction. In the case of selenate, due to the ionic strength dependence of its sorption capacity, the outer-sphere (electrostatic) adsorption mechanism is thought to be the main factor. In complex environments, co-precipitation and surface precipitation are sometimes intermingled with adsorption phenomena.

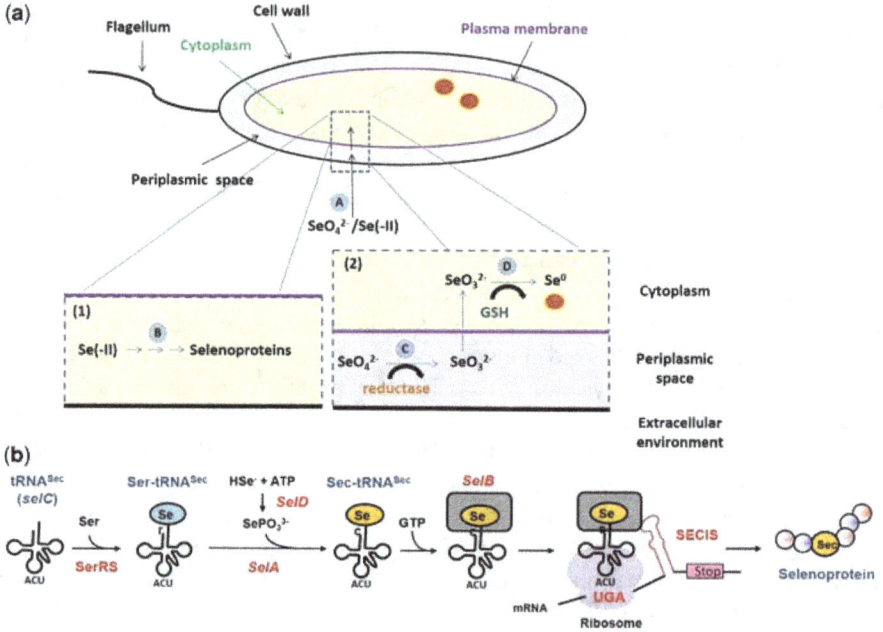

Figure 2.3 Selenium transformations in bacteria related to the supplementation of different forms. a) Assimilatory and dissimilatory metabolism of selenium: (A) Import of various forms (selenate/selenide) of Se inside the bacterial cell; Inset 1 (B): the assimilation of selenide into selenoproteins; Inset 2 (C) respiration of selenate to selenite in G- bacteria; Inset 2 (D) reduction of selenite to red Se^0 by glutathione reductase (GSH). b) Bacterial synthesis of selenoproteins from selenides. (Sec – selenocysteine; *SelA* – selenocysteine synthase; *SelB* – SECIS-binding protein; *SelC* – Sec-specific tRNA (tRNASec); *SelD* – selenophosphate synthetase; Ser – serine; SerRS – seryl-tRNA synthetase; GTP – guanosine-5′-triphosphate; SECIS – selenocysteine insertion sequence) (modified from Staicu *et al.*, 2017).

2.3.3.5 Tungsten

Tungsten (W) is a rare metal that occurs in minerals such as wolframite, (Fe, Mn) WO_4, scheelite, $CaWO_4$, huebnerite, $MnWO_4$, and ferberite, $FeWO_4$ (Lide, 2004). The element reacts with oxygen by forming tungstic oxide, WO_3, which further solubilize in aqueous alkaline solutions and form tungstate ions, WO_4^{2-} (Wiberg & Holleman, 2001). Tungsten supplements are provided to AD reactors in the form of tungstate. From the biological perspective, tungsten is the heaviest element displaying known biological functions. Several oxidoreductase enzymes containing tungsten were characterized and some require selenium for proper functioning (Stiefel, 1998).

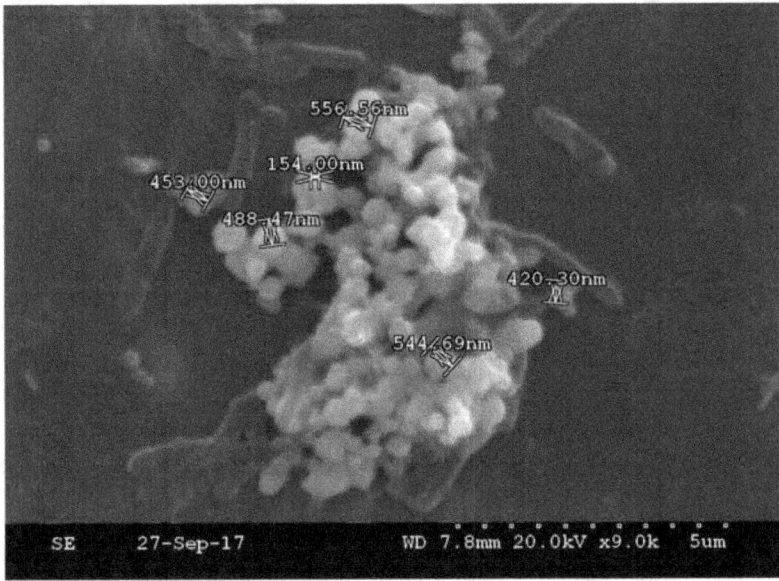

Figure 2.4 Biogenic Se^0 produced by *Sulfospirillum* sp. under anaerobic conditions. Image taken by Scanning Electron Microscopy (SEM) (courtesy of Dr. Gavin Collins and Simon Mills, NUI Galway, Ireland).

Adsorption of tungsten and tetrathiotungstate (WS_4^{2-}) onto pyrite was investigated, showing an increased adsorption capacity with decreasing pH and a higher adsorption of WO_4^{2-} over WS_4^{2-} (Cui & Johannesson, 2016). The authors correlate the adsorption capacity of W forms with different inner-sphere complexation on the pyrite surface. Another study investigated the interaction of tungsten with hematite showing strong adsorption and, similarly to pyrite, increased adsorption capacity with decreasing pH (Rakshit *et al.*, 2017). Tungstate was also shown to adsorb onto organically modified montmorillonite and the main sorption mechanisms involve chemisorption and ion exchange (Muir *et al.*, 2017). The sorption capacity was high at acidic pH (pH < 5), while at pH > 5 sorption was limited or completely inhibited.

2.4 HOW TO ASSESS SPECIATION AND BIOAVAILABILITY OF TE IN ANAEROBIC DIGESTERS

TE can be present under different chemical species (speciation), which can be free or associated to various constituents of the matrix (fractionation). Their determination may be performed using *in situ* or *ex situ* analysis, both approaches having limitations. For example, in the *ex situ* methodologies, inappropriate sampling and/or storage may result in TE sorption/release or redox state modifications. In contrast, *in situ* methodologies are less prone to alteration of the initial speciation and fractionation, but may have lower analytical performances than laboratory techniques.

2.4.1 Sampling

One of the main challenges in TE sampling in anaerobic digesters is the ability to maintain anoxic conditions. Indeed, in anoxic samples, sulfide-based phases may play a significant role in the retention of TE. Upon oxygenation of the sample, these phases can be oxidized, resulting in the release of the associated TE into solution. Among the released metals, Fe and Mn can further generate new solid phases (Fe oxyhydroxides, Mn oxides, etc), which could sorb the TE initially present in the liquid phase or released during the dissolution of solid phases (Almeida *et al.*, 2008; Caetano *et al.*, 2003; Caille *et al.*, 2003). As a result, the concentrations of TE of the original liquid phase can increase or decrease (even below their initial value) according to these antagonistic phenomena. Such evolution can occur within a few hours (Caetano *et al.* 2003). These phenomena may, thus, induce a significant change of the TE partitioning, as highlighted in comparative sequential extractions performed in the presence or absence of air (Buykx *et al.*, 2000; Lenz *et al.*, 2008a). Special care should, thus, be taken during sampling and sample preparation when studying anaerobic systems.

A second challenge lies in the necessity that samples collected from a full-scale biogas reactor should be representative of the whole system. Hence, precautions have to be taken to ensure homogeneity of the reactor material in case of fully mixed system. Sampling may, thus, be the weak point in the analytical procedure as it may represent a major source of uncertainty and bias. Ortner *et al.* (2014) demonstrated that the sampling itself may induce an additional deviation of 6%– 12% at minimum on TE analysis repeatability. It is supposed that this is mainly attributed to inhomogeneities originated from low mixing efficiency in the digesters, which is a typical phenomenon of biogas plants.

2.4.1.1 In-situ sampling

In-situ sampling can be performed through the installation of porous liquid-phase sampling probes in the bioreactor or through the deployment of semipermeable membranes or diffusion-based passive samplers.

Suction-based samplers with hollow fibers (Vink, 2002) or porous ceramic or glass probe heads (Hofacker *et al.*, 2013) can be installed directly in a bioreactor. The samplers deliver a filtered sample by pumping (Duester *et al.*, 2008) or by applying a vacuum (Seeberg-Elverfeldt *et al.*, 2005). The *in situ* bulk/liquid separation is less prone to induce changes in TE partitioning and more straightforward than *ex-situ* separation. The drawbacks are chemical fouling, in particular under strongly varying redox conditions, and biofouling, in particular with high biological activity and long experimental times.

The Donnan Membrane Technique (DMT) is based on semipermeable membranes separating the donor solution (i.e. the anaerobic digester) and an acceptor solution collecting the free metal ions after membrane passage (Temminghoff *et al.*, 2000; Weng *et al.*, 2011). Once the Donnan equilibrium is reach between both solutions, the TE present in the acceptor solution are quantified, generally using a spectrometric technique (e.g. ICP-MS). A main advantage with regard to AD is its applicability to anoxic systems and the avoidance of sample matrix effects. A major limitation arises in the case of low TE presence in the bioreactor, where the concentration in the acceptor solution may be too low to be determined. This technique has already been used to determine Co speciation in anaerobic media (Bartacek *et al.*, 2008).

The diffusive gradients in thin films (DGT) passive sampling is based on the diffusion flux of TE through a diffusive layer (porous gel matrix) followed by its irreversible sorption onto an underlying layer of binding phase (Davison & Zhang, 1994). During the deployment, a TE concentration gradient is established in the diffusive layer between the exposition media and the binding phase. The amount of accumulated TE is thus correlated to its concentration in the exposition media and the duration of deployment. After the deployment, the TE is eluted from the binding phase and quantified, usually by atomic spectrometry, then its original concentration in the exposition media is back-calculated (Davison & Zhang, 1994). Initially developed to assess the sum of free and labile metal concentration in water, the DGT technique has also been used for wastewater, sediments, and soils. In the latter two situations, the accumulation of TE in the DGT device induces a decrease in the TE concentration in the interstitial solution at the vicinity of the device, resulting in some cases in a release of TE from the solid phase. Such behavior could mimic the uptake of TE by plants, making the DGT technique able to estimate the bioavailable TE (Hooda & Zhang, 2008; Sun *et al.*, 2014). In recent studies, the speciation of redox-sensitive TE can also be addressed either by using a binding phase that retains a specific TE species (e.g. a thiolated binding phase will accumulate As (III) only; Bennett *et al.*, 2011) or by analyzing the eluted species with a hyphenated technique (e.g. Hg speciation could be determined by analyzing the DGT eluate by LC-ICP-MS; Cattani *et al.*, 2008). DGT devices main advantages are their ease of use and their ability to limit the risk of modification of the sample by introduction of oxygen. Their major drawback lies in the fact the DGT measurements give only an integrated concentration over the whole deployment duration, and not a dynamic evolution of TE concentration over time. The DGT technique has recently been tested to determine Cd fractionation in biological methane potential tests during anaerobic digestion of dairy waste (Bourven *et al.*, 2017).

2.4.1.2 *Ex-situ sampling*

To avoid oxygenation during *ex-situ* sampling from anaerobic environments (e.g. by pumping slurry samples from a bioreactor), the best practice is probably to directly sample into a glove box with the respective oxygen-free conditions of the reactor. After collection, the sample must be prepared to be compatible with the analytical method. For example, TE should be transferred to a liquid phase to perform atomic spectrometry analysis. In the meantime, sample preservation may also be necessary to limit TE species alteration during transport and storage of the sample, through (de)sorption, chemical or biological processes. Once again, oxygenation should be avoided during all these steps.

One main step of the sample preparation procedure is the separation of solid and liquid phase, which can be carried out by squeezing, centrifugation, sedimentation, filtration or freeze-drying (details on the different methods are available in Bufflap & Allen, 1995a, b). Briefly, squeezing is more useful for work with core sections and less interesting for reactor studies. Filtration is time consuming in comparison to centrifugation, but may suffer from shifts in apparent cut-off due to filter-cake formation. Drying under an inert gas (e.g. N_2) at room temperature or shock freezing by liquid nitrogen with subsequent lyophilization are additional options if analysis of the bulk phase requires dried materials (e.g. Scanning Electron Microscope) (Bordas & Bourg, 1998; Rapin *et al.*, 1986). Centrifugation is more rapid than sedimentation and it seems easier to avoid oxygen contamination if no glove box is available.

Before performing direct solid analysis or extraction of TE, the solid phase needs to be homogenized, which may generally include drying and grinding steps. Such sample preparation methods should be carefully evaluated in order to minimize their influence on original TE speciation and fractionation. Indeed, it has been reported that drying and grinding procedures may affect the results of TE fractionation in anaerobic sediments or biogas slurries (Baeyens *et al.*, 2003; Ortner *et al.*, 2014; Zehl & Einax, 2005). The drying of samples may allow better homogenization compared with wet samples, thus improving the repeatability of sequential extractions (Baeyens *et al.*, 2003). But, if not performed under oxygen-free conditions, it may also result in drastic modification of TE fractionation, either by oxidation of the TE itself or by oxidation of the associated binding phases (Ortner *et al.*, 2014; Zehl & Einax, 2005).

2.4.2 TE speciation in liquid samples

Various analytical techniques can be used to determine the individual species of a given TE in the liquid phase of AD. Most of these species-specific analyses are performed with 'hyphenated techniques', which corresponds to the coupling of a separation technique (typically liquid chromatography, LC, or capillary electrophoresis, CE) with spectrometric detection. Ion exchange chromatography is the most common separation mechanism, followed by reverse phase (usually using an ion pairing agent). Both allow the separation of the main species of, for example, As, Cr, Hg, Sb or Se. For the study of complexes involving TE and organic matter, size exclusion chromatography (SEC) is more suitable (Laborda et al., 2008; Sadi et al., 2002; Vogl & Heumann, 1997). Inductively coupled plasma mass spectrometry (ICP-MS) has become the detector of choice, due to its low limits of detection (a few to sub-μg L^{-1}) for a large range of elements. Several reviews dealing with examples and advances in TE speciation by LC-ICP-MS or CE-ICP-MS can be found in the literature (Harrington et al., 2010, 2015; Harvanova & Bloom, 2015; Michalski et al., 2011; Popp et al., 2010).

The major challenges for TE speciation in AD samples by hyphenated techniques are related to the stability of analytes during the separation step, the complexity of the sample matrix, and the lack of standards for species identification. The fragility of the analyte relates to both complexes and redox stability of the analyte. During the chromatographic separation (at least several minutes), analytes can be oxidized by oxygen dissolved in the eluent. Furthermore, as the eluent required for the separation may not have the same aqueous composition than the original samples, spontaneous precipitation or complexation may occur upon mixing analyte and eluent. This would result in bias due, respectively, to co-precipitation and/or sorption of the analyte on the newly formed phases or to modification of the original complexation equilibrium involving the studied TE. For example, the addition of EDTA in the mobile phase may induce a dissociation of weak complexes of antimony (Sb) and organic acids (Hansen et al., 2011). Mobile-phase composition and pH should thus be compatible with the AD sample. Once the different species are correctly separated, they can be individually identified by matching retention times with known standards and then quantified. Unfortunately, such standards are not always available, in particular for elements that are known to readily form soluble species with S, such as Se and As (Lenz et al., 2008b; Petrov et al., 2012; Planer-Friedrich et al., 2010). Due to the low concentration of the analytes, the identification by collection of chromatographic fractions and subsequent analysis using mass spectrometry is not a straightforward option.

2.4.3 TE analysis in solid samples

2.4.3.1 TE speciation

To be able to conduct species-specific spectroscopic measures in non-crystalline complex matrices with low concentration of analyte, the high brilliance of synchrotron radiation (SR) is a prerequisite for success (Table 2.3). There are currently about 50 SR light sources around the world (www.lightsources.org), but so far there are relatively few reports on the use of SR to characterize the chemical speciation of samples in AD. Due to the low abundance of TE in AD, often below μg kg^{-1}, speciation remains generally inaccessible for spectroscopic methods. However, the on-going development of SR sources with increasing light brilliance is continually lowering the detection limits. Today, Fe is the only metal that, at least in some AD samples, reaches concentrations that can be directly studied by use of Extended X-ray Absorption Fine Structure (EXAFS) spectroscopy in bulk samples.

Table 2.3 Synchrotron-radiation spectroscopic methods available for TE and major elements speciation in samples from AD reactors.

Methods	Advantages	Drawbacks
XANES (NEXAFS)	Can be used to separate classes of mainly low mass elements	Provide more fingerprinting information (mostly redox state) than strict species-specificity
EXAFS	A non-destructive method in complex matrices that provides species-specific information	The sensitivity is restricted to about $1-10$ $\mu g\ g^{-1}$ for the most brilliant 4th generation SR sources. Radiation damage is a concern
μ-XANES, μ-EXAFS	Provides XANES/EXAFS species-specific information at micro-scale	Similar to conventional XANES/EXAFS, but higher energy flux required
STXM	Localized (nm scale) species-specific (XANES) information in contrasting images	Sub nm thickness of samples required making the method less sensitive to dispersed elements. For TE high local concentrations is required (e.g. nano-materials)
XPS	Surface-sensitive method for chemical speciation within the nm scale of surfaces	TE addition is required due to insensitivity

Similar to organic soils and sediments, the matrix of AD samples is dominated by low mass elements such as H, C, N, O, and S. Biological and physical structures built by these elements can be visualized in high detail by, for instance, confocal microscopy, and transmission and scanning-electron microscopy (TEM and SEM, respectively). Coupling the latter with energy-dispersive X-ray absorption (EDX), TEM/SEM provides useful tools for elemental mapping linking structure to the occurrence of TE. The recent development of SR techniques opens up the possibility for species-specific identification of most elements in the periodic table.

For Fe, in principal species-specific information can be provided by K-edge X-ray Absorption Near-Edge Spectroscopy (XANES) (O'Day *et al.*, 2004). Because S and Fe react with each other and form metastable and stable chemical compounds such as FeS and FeS_2, a combination of S and Fe XANES is highly beneficial for the characterization of anoxic environments. In this way the uncertainty, which is in the order of 15% in the quantification of S and Fe species by fitting model compounds to their respective XANES spectra, could be lowered substantially. At SR beamlines with focusing lenses, XANES spectra for C, N, S, and Fe can be provided with high spatial resolution (μ-XANES).

A technique that partly avoids the problem with low TE concentration samples is Scanning Transmission X-ray Microscopy (STXM). This technique is primarily oriented towards the low mass elements, in particular C, and provides high-quality electron-density microscope images in essence similar to SEM/TEM (Hitchcock et al., 2005; Kinyangi et al., 2006). Because of the high resolution of this technique, localized high concentrations of heavier elements (e.g. TE) can be detected and its coordination chemistry unraveled. Thus, nano-particles of TE (e.g. metal sulfides) or local concentrations of TE at e.g. cell surfaces or in cell vacuoles can be identified by STXM (Behrens et al., 2012). One problem is that samples need to be very thin to enable transmission detection. Therefore, specific sample holder cells and detectors sensitive to TE fluorescence currently are under development at beamlines devoted to STXM.

The use of more conventional and more accessible spectroscopic techniques, other than synchrotron-based techniques in the characterization of TE speciation also offers important information. For example, Electron Paramagnetic Resonance (EPR) is very sensitive with fluorescence allowing analysis of very low TE concentrations, yielding information about TE oxidation state, geometry, and chemical environment of e.g. for Cu, Fe(II), and Fe(III). Mössbauer Spectroscopy (for Fe) can provide information on the nature of major solid phases, but also about the oxidation state of TE. Although these techniques are only useful for certain types of TE, they can contribute significantly to mechanistic knowledge on TE fate.

2.4.3.2 TE fractionation using sequential extraction

To assess TE fractionation, which can provide information regarding their potential mobility and bioavailability in AD ecosystems, different chemical extraction approaches might be used. However, the results obtained are operationally defined, i.e. the 'forms' of TE are a direct result of the used extraction procedure (Quevauviller et al., 1997).

The sequential extraction (SE) approach lies on successive extractions, involving a series of reagents selected for their ability to react with different major solid components of the sample, to release the associated TE. This method has been popularized by Tessier et al. (1979) who developed an analytical procedure to differentiate particulate cationic TE (Cd, Co, Cu, Ni, Pb, Zn, Fe, and Mn) into five main fractions: exchangeable, bound to carbonates, bound to Fe–Mn oxides, bound to organic matter, and residual. Its application to anaerobically treated sludge revealed that organic matter and sulfide fractions are the most important carriers of metals in these matrices (Angelidis & Gibbs, 1989), but both fractions are simultaneously extracted and therefore no information regarding the contribution of each phase in TE binding is provided. However, before the publication of the Tessier procedure, Stover *et al.* (1976) developed a SE scheme for determining the fractionation of Pb, Zn, Cu, and Ni in anaerobically digested wastewater sludges from municipal treatment plants. The Stover scheme discriminates the TE present in exchangeable, adsorbed, organic matter-bound, carbonate, and sulfide forms. This, therefore, allows discriminating metal distribution between the organic matter and sulfide fractions. However, the higher number of extraction steps compared with the Tessier scheme results in a poor recovery of the TE extracted compared with the initial TE budget (van Hullebusch *et al.*, 2005). The Bureau Communautaire de Reference (BCR) developed a SE scheme to harmonize all SE studies and certified SE through inter-laboratory trials (Quevauviller, 1994). An accelerated BCR scheme, based on ultrasound-assisted extractions, was also proposed to significantly reduce the extraction time: from 2.5 days for Tessier scheme to half a day (Perez-Cid *et al.*, 1999).

Even though several SE schemes have been proposed, it must be mentioned that no single fractionation scheme solves for distinct TE bearing phases exclusively and exhaustively (Filgueiras *et al.*, 2002). Nevertheless, despite uncertainties in the selectivity of the various extractants and possible problems due to re-adsorption and partial oxidation of oxygen sensitive elements (e.g. Fe and S), SE procedures are a well established, justified means to study metal partitioning among the various solid phases of soils and sludges (see Filgueiras *et al.*, 2002 for overview).

There are also approaches to adapt existing SE procedures for improved study bioavailability of anions such as Se or Mo (Wright *et al.*, 2003). However, Lenz *et al.* (2008b) showed that the interpretation can be biased by unselective extraction of targeted species and artefacts introduced during the extraction, as later discussed by Huang and Kretzschmar (2010) for As chemical species.

Shakeri Yekta *et al.* (2012) investigated the effect of SE of TE on S speciation in anoxic sludge samples from two lab-scale biogas reactors augmented with Fe. Analyses of S K-edge XANES spectroscopy and Acid Volatile Sulfides (AVS) were conducted on the residues from each SE step. The S speciation in sludge samples after AVS analysis was also determined. In the anoxic solid phase, S was mainly present as FeS (~60% of total S) and reduced organic S (~30% of total S), such as organic sulfide and thiol groups. During the first step of the extraction procedure (the removal of exchangeable cations), a part of the FeS fraction corresponding to 20% of total S was transformed to zero-valent S. Fe was not released into the solution during this extraction step. After the last extraction step (organic/sulfide fraction) a secondary Fe phase was formed. The change in chemical speciation of S and Fe occurring during SE procedure suggests indirect effects on TE associated to the FeS fraction that may lead to incorrect results. Furthermore, the FeS fraction was quantitatively dissolved by AVS extraction. These results identify critical limitations for the application of SE for TE speciation analysis outside the framework for which the methods were developed.

2.5 CONCLUDING REMARKS

Trace elements are often supplemented in high amounts. This is mainly due to the lack of clear understanding of TE speciation and bioavailability. Initially, the amount of added TE was generally determined based on the total rather than 'free' or bioavailable TE concentrations. However, it is now clear that chemical speciation is critical when predicting bioavailability. Although TE chemical fractionation has been studied extensively over the last two decades to determine the fate of TE in anaerobic environments, its relationship with TE bioavailability is still not clearly understood. Such studies are further complicated as non-redox sensitive (e.g. Co, Cu, Ni, Zn) and redox-sensitive (e.g. Fe, Mn, Mo, Se, W) TE are involved in different biogeochemical reactions with mineral and organic phases present in AD.

The well known chemical sequential extraction methods and, more recently, some more advanced analytical techniques such as DMT or XAS have been applied to determine the speciation of TE in liquid and solid samples, respectively; however, the application of the advanced analytical techniques to anaerobic digesters is still in development.

In conclusion, a fundamental approach, which includes the basic biogeochemical processes of TE and the available analytical techniques, is highly needed. From a practical point of view, a deeper understanding of TE bioavailability in anaerobic digestion will help to minimize TE supplementation costs, while also maximizing the performance of anaerobic treatment processes operated at full scale.

ACKNOWLEDGEMENTS

The authors gratefully acknowledge EU COST Action ES1302 'Ecological functions of trace metals in anaerobic biotechnologies' for making this collaborative project possible.

REFERENCES

Almeida C. M. R., Mucha A. P., Bordalo A. A. and Vasconcelos M. T. S. D. (2008). Influence of a salt marsh plant (*Halimione portulacoides*) on the concentrations and potential mobility of metals in sediments. *Science of The Total Environment*, **403**, 188–195.

Angelidis M. and Gibbs R. J. (1989). Chemistry of metals in anaerobically treated sludges. *Water Research*, **23**, 29–33.

Baeyens W., Monteny F., Leermakers M. and Bouillon S. (2003). Evaluation of sequential extractions on dry and wet sediments. *Analytical and Bioanalytical Chemistry*, **376**, 890–901.

Banks C. J., Zhang Y., Jiang Y. and Heaven S. (2012). Trace element requirements for stable food waste digestion at elevated ammonia concentrations. *Bioresource Technology*, **104**, 127–135.

Bartacek J., Fermoso F. G., Baldo-Urrutia A. M., van Hullebusch E. D. and Lens P. N. L. (2008). Co toxicity in anaerobic granular sludge: Influence of chemical speciation. *Journal of Industrial Microbiology and Biotechnology*, **35**, 1465–1474.

Behrens S., Kappler A. and Obst M. (2012). Linking environmental processes to the in situ functioning of microorganisms by high-resolution secondary ion mass spectrometry (Nano-SIMS) and scanning transmission X-ray microscopy (STXM). *Environmental Microbiology*, **14**, 2851–2869.

Bennett W. W., Teasdale P. R., Panther J. G., Welsh D. T. and Jolley D. F. (2011). Speciation of dissolved inorganic arsenic by diffusive gradients in thin films: Selective binding of As^{III} by 3-mercaptopropyl-functionalized silica gel. *Analytical Chemistry*, **83**, 8293–8299.

Bischofsberger W., Dichtl N., Rosenwinkel K.-H., Seyfried C. F. and Böhnke B. (2005). *Anaerobtechnik*, 2nd rev. ed. Springer-Verlag, Berlin.

Bordas F. and Bourg A. C. M. (1998). A critical evaluation of sample pretreatment for storage of contaminated sediments to be investigated for the potential mobility of their heavy metal load. *Water, Air & Soil Pollution*, **103**, 137–149.

Bourven I., Casellas M., Buzier R., Lesieur J., Lenain J.-F., Faix A., Bressolier P., Maftah C. and Guibaud G. (2017). Potential of DGT in a new fractionation approach for studying trace metal element impact on anaerobic digestion: The example of cadmium. *International Biodeterioration & Biodegradation*, **119**, 188–195.

Braud A., Hoegy F., Jezequel K., Lebeau T. and Schalk I. J. (2009). New insights into the metal specificity of the *Pseudomonas aeruginosa* pyoverdine-iron uptake pathway. *Environmental Microbiology*, **11**, 1079–1091.

Bruggeman C., Maes A. and Vancluysen J. (2007). The interaction of dissolved Boom Clay and Gorleben humic substances with selenium oxyanions (selenite and selenate). *Applied Geochemistry*, **22**, 1371–1379.

Bufflap S. E. and Allen H. E. (1995a). Comparison of pore water sampling techniques for trace metals. *Water Research*, **29**, 2051–2054.

Bufflap S. E. and Allen H. E. (1995b). Sediment pore water collection methods for trace metal analysis: A review. *Water Research*, **2**, 165–177.

Buykx S. E. J., Bleijenberg M., Van den Hoop M. A. G. T. and Loch J. P. G. (2000). The effect of oxidation and acidification on the speciation of heavy metals in sulfide-rich freshwater sediments using a sequential extraction procedure. *Journal of Environmental Monitoring*, **2**, 23–27.

Caetano M., Madureira M. J. and Vale C. (2003). Metal remobilisation during resuspension of anoxic contaminated sediment: Short-term laboratory study. *Water, Air & Soil Pollution*. **143**, 23–40.

Caille N., Tiffreau C., Leyval C. and Morel J. L. (2003). Solubility of metals in an anoxic sediment during prolonged aeration. *Science of The Total Environment*, **301**, 239–250.

Cattani I., Spalla S., Beone G. M., Del Re A. A. M., Boccelli R. and Trevisan M. (2008). Characterization of mercury species in soils by HPLC-ICP-MS and measurement of fraction removed by diffusive gradient in thin films. *Talanta*, **74**, 1520–1526.

Cordoba P. and Staicu L. C. (2018). Flue Gas Desulfurization effluents: An unexploited selenium resource. *Fuel*, **223**, 268–276.

Cui M. and Johannesson K. H. (2016). Comparison of tungstate and tetrathiotungstate adsorption onto pyrite. *Chemical Geology*, **464**, 57–68.

Davison W. and Zhang H. (1994). In situ speciation measurements of trace components in natural waters using thin-film gels. *Nature*, **367**, 546–548.

Demirel B. and Scherer P. (2011). Trace element requirements of agricultural biogas digesters during biological conversion of renewable biomass to methane. *Biomass & Bioenergy*, **35**, 992–998.

Duester L., Vink J. P. M. and Hirner A. (2008). Methylantimony and arsenic species in sediment pore water tested with the sediment or fauna incubation experiment. *Environmental Science & Technology*, **42**, 5866–5871.

Fermoso F. G., Bartacek J., Chung L. C. and Lens P. N. L. (2008). Supplementation of Co to UASB reactors by pulse dosing: $CoCl_2$ versus $CoEDTA^{2-}$ pulses. *Biochemical Engineering Journal*, **42**, 111–119.

Fermoso F. G., Bartacek J., Jansen S. and Lens P. N. L. (2009). Metal supplementation to UASB bioreactors: From cell-metal interactions to full-scale application. *Science of the Total Environmen*, **407**, 3652–3667.

Fermoso F. G., Bartacek J., Manzano R., van Leeuwen H. P. and Lens P. N. L. (2010). Dosing of anaerobic granular sludge bioreactors with Co: Impact of Co retention on methanogenic activity. *Bioresource Technology*, **101**, 9429–9437.

Fermoso F. G., Van Hullebusch E. D., Guibaud G., Collins G., Svensson B. H., Carliell-Marquet C., Vink J. P. M., Esposito G. and Frunzo L. (2015). Fate of trace metals in anaerobic digestion. *Advances in Biochemical Engineering/Biotechnology*, **151**, 171–195.

Fernandez-Martinez A. and Charlet L. (2009). Selenium environmental cycling and bioavailability: A structural chemist point of view. *Reviews in Environmental Science and Bio/Technology*, **8**, 81–110.

Filgueiras A. V., Lavilla I. and Bendicho C. (2002). Chemical sequential extraction for metal partitioning in environmental solid samples. *Journal of Environmental Monitoring*, **4**, 823–857.

Glass J. B. and Orphan V. J. (2012). Trace metal requirements for microbial enzymes involved in the production and consumption of methane and nitrous oxide. *Frontiers in Microbiology*, **3**, 61.

Gustavsson J., Svensson B. H. and Karlsson A. (2011). The feasibility of trace element supplementation for stable operation of wheat stillage-fed biogas tank reactors. *Water Science and Technology*, **64**, 320–325.

Gustavsson J., Shakeri Yekta S., Sundberg C., Karlsson A., Ejlertsson J., Skyllberg U. and Svensson B. H. (2013a). Bioavailability of Co and nickel during anaerobic digestion of sulfur-rich stillage for biogas formation. *Applied Energy*, **112**, 473–477.

Gustavsson J., Yekta S. S., Karlsson A., Skyllberg U. and Svensson B. H. (2013b). Potential bioavailability and chemical forms of Co and Ni in the biogas process-An evaluation based on sequential and acid volatile sulfide extractions. *Engineering in Life Sciences*, **13**(6), 572–579.

Hansen C., Schmidt B., Larsen E. H., Gammelgaard B., Sturup S. and Hansen H. R. (2011). Quantitative HPLC-ICP-MS analysis of antimony redox speciation in complex sample matrices: New insights into the Sb-chemistry causing poor chromatographic recoveries. *Analyst*, **136**, 996–1002.

Harrington C. F., Vidler D. S. and Jenkins R. O. (2010). Analysis of organometal(loid) compounds in environmental and biological samples. *Metal Ions in Life Sciences*, **7**, 33–69.

Harrington C. F., Clough R., Hill S. J., Madrid Y. and Tyson J. F. (2015). Atomic spectrometry update: Review of advances in elemental speciation. *Journal of Analytical Atomic Spectrometry*, **30**, 1427–1468.

Harvanova J. and Bloom L. (2015). Capillary electrophoresis technique for metal species determination: A review. *Journal of Liquid Chromatography & Related Technologies*, **38**, 371–380.

Hitchcock A. P., Morin C., Zhang X., Araki T., Dynes J. J., Stover H., Brash J. L., Lawrence J. R. and Leppard G. G. (2005). Soft X-ray spectromicroscopy of biological and synthetic polymer systems. *Journal of Electron Spectroscopy and Related Phenomena*, **144–147**, 259–269.

Hofacker A. F., Voegelin A., Kaegi R. and Kretzschmar R. (2013). Mercury mobilization in a flooded soil by incorporation into metallic copper and metal sulfide nanoparticles. *Environmental Science & Technology*, **47**, 7739–7746.

Hooda P. S. and Zhang H. (2008). DGT measurements to predict metal bioavailability in soils. *Developments in Soil Science*, **32**, 169–185.

Huang J. H. and Kretzschmar R. (2010). Sequential extraction method for speciation of arsenate and arsenite in mineral soils. *Analytical Chemistry*, **82**, 5534–5540.

Jarvis Å., Nordberg Å., Jarlsvik T., Mathisen B. and Svensson B. H. (1997). Improvement of a grass-clover silage-fed biogas process by the addition of Co. *Biomass and Bioenergy*, **12**, 453–460.

Kinyangi J., Solomon D., Liang B., Lerotic M., Wirick S. and Lehmann J. (2006). Nanoscale biogeocomplexity of the organomineral assemblage in soil: Application of STXM microscopy and C 1s-NEXAFS spectroscopy. *Soil Science Society of America Journal*, **70**, 1708–1718.

Kloss R. (1986). Planung von Biogasanlagen nach technisch-wirtschaftlichen Kriterien: Stand des Wissens, Stand der Technik in der Landwirtschaft; Planungsmodell; mit Hinweisen für die Planung abwassertechnischer Anlagen. Oldenbourg. Available online: https://docplayer.org/73464624-Planung-landwirtschaftlicher-biogasanlagen-nach-technisch-wirtschaftlichen-kriterien.html (In German)

Kyle J. H., Breuer P. L., Bunney K. G., Pleysier R. and May P. M. (2011). Review of trace toxic elements (Pb, Cd, Hg, As, Sb, Bi, Se, Te) and their deportment in gold processing. Part 1: Mineralogy, aqueous solution chemistry and toxicity. *Hydrometallurgy*, **107**, 91–100.

Laborda F., Bolea E., Gorriz M. P., Martin-Ruiz M. P., Ruiz-Begueria S. and Castillo J. R. (2008). A speciation methodology to study the contributions of humic-like and fulvic-like acids to the mobilization of metals from compost using size exclusion chromatography-ultraviolet absorption-inductively coupled plasma mass spectrometry and deconvolution analysis. *Analytica Chimica Acta*, **606**, 1–8.

Labunskyy V. M., Hatfield D. L. and Gladyshev V. N. (2014). Selenoproteins: Molecular pathways and physiological roles. *Physiological Reviews*, **94**, 739–777.

Le Bars M., Legros S., Levard C., Chaurand P., Tella M., Rovezzi M., Browne P., Rose J. and Doelsch E. (2018). Drastic change in zinc speciation during anaerobic digestion and composting: Instability of nanosized zinc sulfide. *Environmental Science & Technology*, **52**, 12987–12996.

Legros S., Levard C., Marcato-Romain C.-E., Guiresse M. and Doelsch E. (2017). Anaerobic Digestion Alters Copper and Zinc Speciation. *Environmental Science & Technology*, **51**, 10326–10334.

Lenz M., Hullebusch E. D. V, Hommes G., Corvini P. F. X. and Lens P. N. L. (2008a). Selenate removal in methanogenic and sulfate-reducing upflow anaerobic sludge bed reactors. *Water Research*, **42**, 2184–2194.

Lenz M., Van Hullebusch E. D., Farges F., Nikitenko S., Borca C. N., Grolimund D. and Lens P. N. L. (2008b). Selenium speciation assessed by X-ray absorption spectroscopy of sequentially extracted anaerobic biofilms. *Environmental Science & Technology*, **42**, 7587–7593.

Lide D. R. (2004). CRC Handbook of Chemistry and Physics, 84th edn. CRC Press, Boca Raton.

Maharaj B. C., Mattei M. R., Frunzo L., van Hullebusch E. D. and Esposito G. (2018). ADM1 based mathematical model of trace element precipitation/dissolution in anaerobic digestion processes. *Bioresource Technology*, **267**, 666–676.

Marcato C. E., Pinelli E., Cecchi M., Winterton P. and Guiresse M. (2009). Bioavailability of Cu and Zn in raw and anaerobically digested pig slurry. *Ecotoxicology and Environmental Safety*, **72**(5), 1538–1544.

Marie Sych J., Lacroix C. and Stevens M. J. (2016). Vitamin B_{12} – Physiology, Production and Application. In: Industrial Biotechnology of Vitamins, Biopigments, and Antioxidants, E. J. Vandamme and J. L. Revuelta (eds), Wiley, New York, pp. 129–160.

Matthews R. G. (2001). Cobalamin-Dependent Methyltransferases. *Accounts of Chemical Research*, **34**(8), 681–689.

Michalski R., Jabłonska M., Szopa S. and Lyko A. (2011). Application of ion chromatography with ICP-MS or MS detection to the determination of selected halides and metal/metalloids species. *Critical Reviews in Analytical Chemistry*, **41**, 133–150.

Moestedt J., Nordell E., Shakeri Yekta S., Lundgren J., Martí M., Sundberg C., Ejlertsson J., Svensson B. H. and Björn A. (2015). Effects of trace element addition on process stability during anaerobic co-digestion of OFMSW and slaughterhouse waste. *Waste Management*, **8**, 2600–2621.

Molaey R., Bayrakdar A. and Calli B. (2018a). Long-term influence of trace element deficiency on anaerobic mono-digestion of chicken manure. *Journal of Environmental Management*, **223**, 743–748.

Molaey R., Bayrakdar A., Surmeli R. O. and Calli B. (2018b). Influence of trace element supplementation on anaerobic digestion of chicken manure: Linking process stability to methanogenic population dynamics. *Journal of Cleaner Production*, **181**, 794–800.

Monlau F., Sambusiti C., Ficara E., Aboulkas A., Barakat A. and Carrère H. (2015). New opportunities for agricultural digestate valorization: Current situation and perspectives. *Energy and Environmental Science*, **8**, 2600–2621.

Möller K. and Müller T. (2012). Effects of anaerobic digestion on digestate nutrient availability and crop growth: A review. *Engineering in Life Sciences*, **12**(3), 242–257.

Mudrack K. and Kunst S. (2003). Biologie der Abwasserreinigung, 5th ed. Spektrum Akademischer Verlag, Berlin. (In German)

Muir B., Andrunik D., Hyla J. and Bajda T. (2017). The removal of molybdates and tungstates from aqueous solution by organo-smectites. *Applied Clay Science*, **136**, 8–17.

Neilands J. B. (1995). Siderophores: Structure and function of microbial iron transport compounds. *Journal of Biological Chemistry*, **270**, 26723–26726.

NRC (2003). Bioavailability of Contaminants in Soils and Sediments Processes, Tools, and Applications. Committee on Bioavailability of Contaminants in Soils and Sediments. National Research Council of the National Academies. The National Academies Press, Washington, D.C.

O'Day P. A., Rivera N. Jr., Root R. and Carroll S. A. (2004). X-ray absorption spectroscopic study of Fe reference compounds for the analysis of natural sediments. *American Mineralogist*, **89**, 572–585.

Ortner M., Rachbauer L., Somitsch W. and Fuchs W. (2014). Can bioavailability of trace nutrients be measured in anaerobic digestion? *Applied Energy*, **126**, 190–198.

Perez-Cid B., Lavilla I. and Bendicho C. (1999). Comparison between conventional and ultrasound accelerated Tessier sequential extraction schemes for metal fractionation in sewage sludge. *Fresenius' Journal of Analytical Chemistry*, **363**, 667–672.

Petrov P. K., Charters J. W. and Wallschlager D. (2012). Identification and determination of selenosulfate and selenocyanate in flue gas desulfurization waters. *Environmental Science & Technology*, **46**, 1716–1723.

Pinto-Ibieta F., Serrano A., Jeison D., Borja R. and Fermoso F. G. (2016). Effect of Co supplementation and fractionation on the biological response in the biomethanization of Olive Mill Solid Waste. *Bioresource Technology*, **211**, 58–64.

Pitter P. (2009). Hydrochemie (Hydrochemistry), Vydavatelství VŠCHT Praha, Prague, Czech Republic. (In Czech)

Planer-Friedrich B., Suess E., Scheinost A. C. and Wallschlager D. (2010). Arsenic speciation in sulfidic waters: Reconciling contradictory spectroscopic and chromatographic evidence. *Analytical Chemistry*, **82**, 10228–10235.

Pobeheim H., Munk B., Johansson J. and Gruebitz G. M. (2010). Influence of trace elements on methane formation from a synthetic model substrate for maize silage. *Bioresource Technology*, **101**(2), 836–839.

Popp M., Hann S. and Koellensperger G. (2010). Environmental application of elemental speciation analysis based on liquid or gas chromatography hyphenated to inductively coupled plasma mass spectrometry – A review. *Analytical Chimica Acta*, **668**, 114–129.

Quevauviller P., Rauret G., Muntau H., Ure A. M., Rubio R., López-Sánchez J. F., Fiedler H. D. and Griepink B. (1994). Evaluation of a sequential extraction procedure for the determination of extractable trace metal contents in sediments. *Fresenius' Journal of Analytical Chemistry*, **349**, 808–814.

Quevauviller P., Rauret G., Lopez-Sanchez J. F., Rubio R., Ure A. and Muntau H. (1997). Certification of trace metal extractable contents in a sediment reference material (CRM 601) following a three-step sequential extraction procedure. *Science of The Total Environment*, **205**, 223–234.

Rakshit S., Sallman B., Davantes A. and Lefevre G. (2017). Tungstate (VI) sorption on hematite: An *in situ* ATR-FTIR probe on the mechanism. *Chemosphere*, **168**, 685–691.

Rapin F., Teissier A., Campbell P. G. C. and Carignan R. (1986). Potential artifacts in the determination of metal partitioning in sediments by a sequential extraction procedure. *Environmental Science & Technology*, **20**, 836–840.

Rempel S., Colucci E., de Gier J. W., Guskov A. and Slotboom D. J. (2018). Cysteine-mediated decyanation of vitamin B_{12} by the predicted membrane transporter BtuM. *Nature Communications*, **9**, 3038.

Romero-Guiza M. S., Vila J., Mata-Alvarez J., Chimenos J. M. and Astals S. (2016). The role of additives on anaerobic digestion: A review. *Renewable & Sustainable Energy Reviews*, **58**, 1486–1499.

Sadi B. B. M., Wrobel K., Kannamkumarath S. S., Castillo J. R. and Caruso J. A. (2002). SEC-ICP-MS studies for elements binding to different molecular weight fractions of humic substances in compost extract obtained from urban solid waste. *Journal of Environmental Monitoring*, **4**, 1010–1016.

Sahm H. (1981). Biologie der Methan-Bildung. *Chemie Ingenieur Technik*, **53**, 854–863. (In German)

Seeberg-Elverfeldt J., Schlüter M., Feseker T. and Kölling M. (2005). Rhizon sampling of pore waters near the sediment/water interface of aquatic systems. *Limnology and Oceanography: Methods*, **3**, 361–371.

Schattauer A., Abdoun E., Weiland P., Plöchl M. and Heiermann M. (2011). Abundance of trace elements in demonstration biogas plants. *Biosystems Engineering*, **108**, 57–65.

Schmidt T., Nelles M., Scholwin F. and Pröter J. (2014). Trace element supplementation in the biogas production from wheat stillage – Optimization of metal dosing. *Bioresource Technology*, **168**, 80–85.

Schnürer A. (2016). Biogas production: Microbiology and technology. In: Advances in Biochemical EngineeringBiotechnology, R. Hatti-Kaul, G. Mamo and B. Mattiasson (eds), Springer, Cham, Vol. **156**, pp. 195–234.

Seyfried C. F., Bode H., Austermann-Haun U., Brunner G., von Hagel G. and Kroiss H. (1990). Anaerobe Verfahren zur Behandlung von Industrieabw€assern. *Korrespondenz Abwasser*, **37**, 1247–1251. (In German)

Shakeri Yekta S., Gustavsson J., Svensson B. H. and Skyllberg U. (2012). Sulfur K-edge XANES and acid volatile sulfide analyses of changes in chemical speciation of S and Fe during sequential extraction of trace metals in anoxic sludge from biogas reactors. *Talanta*, **89**, 470–477.

Shakeri Yekta S., Svensson B. H., Björn A. and Skyllberg U. (2014). Thermodynamic modeling of iron and trace metal solubility and speciation under sulfidic and ferruginous conditions in full scale continuous stirred tank biogas reactors. *Applied Geochemistry*, **47**, 61–73.

Staicu L. C. and Barton L. L. (2017). Microbial metabolism of selenium – for survival or profit. In: Bioremediation of Selenium Contaminated Wastewaters, E. D. van Hullebusch (ed.), Springer, Berlin, pp. 1–31.

Staicu L. C., Oremland R. S., Tobe R. and Mihara H. (2017). Bacteria vs. selenium: A view from the inside out. In: Selenium in Plants, E. A. H. Pilon-Smits, L. Winkel and Z. Q. Lin (eds), Springer, Berlin, pp. 79–108.

Stiefel E. I. (1998). Transition metal sulfur chemistry and its relevance to molybdenum and tungsten enzymes. *Pure and Applied Chemistry*, **70**, 889–896.

Stover R. C., Sommers L. E. and Silviera D. J. (1976). Evaluation of metals in wastewater sludge. *Journal of the Water Pollution Control Federation*, **48**, 2165–2175.

Sun Q., Chen J., Ding S., Yao Y. and Chen Y. (2014). Comparison of diffusive gradients in thin film technique with traditional methods for evaluation of zinc bioavailability in soils. *Environmental Monitoring and Assessment*, **186**, 6553–6564.

Takashima M., Speece R. E. and Parkin G. F. (1990). Mineral requirements for methane fermentation. *Critical Reviews in Environmental Control*, **19**, 465–479.

Temminghoff E. J. M., Plette A. C. C., Van Eck R. and Van Riemsdijk W. H. (2000). Determination of the chemical speciation of trace metals in aqueous systems by the Wageningen Donnan membrane technique. *Analytica Chimica Acta*, **417**, 149–157.

Tessier A., Campbell P. G. C. and Bisson M. (1979). Sequential extraction procedure for the speciation of particulate trace metals. *Analytical Chemistry*, **51**, 844.

Thanh P. M., Ketheesan B., Yan Z. and Stuckey D. (2016). Trace metal speciation and bioavailability in anaerobic digestion: A review. *Biotechnology Advances*, **34**, 122–136.

van Hullebusch E. D., Zandvoort M. H. and Lens P. N. L. (2003). Metal immobilisation by biofilms: Mechanisms and analytical tools. *Reviews in Environmental Science and Biotechnology*, **2**, 9–33.

van Hullebusch E. D., Utomo S., Zandvoort M. H. and Lens P. N. L. (2005). Comparison of three sequential extraction procedures to describe metal fractionation in anaerobic granular sludges. *Talanta*, **65**, 549–558.

van Hullebusch E. D., Guibaud G., Simon S., Lenz M., Yekta S. S., Fermoso F. G., Jain R., Duester L., Roussel J., Guillon E., Skyllberg U., Almeida C. M. R., Pechaud Y., Garuti M., Frunzo L., Esposito G., Carliell-Marquet C., Ortner M. and Collins G. (2016). Methodological approaches for fractionation and speciation to estimate trace element bioavailability in engineered anaerobic digestion ecosystems: An overview. *Critical Reviews in Environmental Science and Technology*, **46**, 1324–1366.

Vink J. P. M. (2002). Measurement of heavy metal speciation over redox gradients in natural water-sediment interfaces and implications for uptake by benthic organisms. *Environmental Science & Technology*, **36**, 5130–5138.

Vogl J. and Heumann K. G. (1997). Determination of heavy metal complexes with humic substances by HPLC/ICP-MS coupling using on-line isotope dilution technique. *Fresenius' Journal of Analytical Chemistry*, **359**, 438–441.

Weiland P. (2006). Anforderungen an Pflanzen seitens des Biogasanlagenbetreibers, TLL-Jena, Eigenverlag. *Thüringer Bioenergietag*, **12**, 26–32. (In German)

Weng L., Alonso Vega F. and Van Riemsdijk W. H. (2011). Strategies in the application of the Donnan membrane technique. *Environmental Chemistry*, **8**, 466–474.

Wiberg E. and Holleman A. F. (2001). Inorganic Chemistry. Academic Press, New York.

Wichard T., Bellenger J. P., Morel F. M. M. and Kraepiel A. M. (2009). Role of the siderophore azotobactin in the bacterial acquisition of nitrogenase metal cofactors. *Environmental Science & Technology*, **43**, 7218–7224.

Wright M. T., Parker D. R. and Amrhein C. (2003). Critical evaluation of the ability of sequential extraction procedures to quantify discrete forms of selenium in sediments and soils. *Environmental Science & Technology*, **37**, 4709–4716.

Xu N., Christodoulatos C. and Braida W. (2005). Adsorption of molybdate and tetrathiomolybdate onto pyrite and goethite: Effect of pH and competitive anions. *Chemosphere*, **62**, 1726–1735.

Yekta S. S., Skyllberg U., Danielsson Å., Björn A. and Svensson B. H. (2017). Chemical speciation of sulfur and metals in biogas reactors – Implications for Co and nickel bio-uptake processes. *Journal of Hazardous Materials*, **324**, 110–116.

Zehl K. and Einax J. W. (2005). Influence of atmospheric oxygen on heavy metal mobility in sediment and Soil. *Journal of Soils and Sediments*, **5**, 164–170.

Zhang J., Zhang Y., Quan X., Liu Y., An X., Chen S. and Zhao H. (2011). Bioaugmentation and functional partitioning in a zerovalent iron-anaerobic reactor for sulfate-containing wastewater treatment. *Chemical Engineering Journal*, **174**, 159–165.

Zhang Y. and Gladyshev V. N. (2009). Comparative genomics of trace elements: Emerging dynamic view of trace element utilization and function. *Chemical Reviews*, **109**, 4828–4861.

Anaerobic Digestion and Important Trace Element Enzymes

Juan M. Gonzalez[1] and Blaz Stres[2,3,4]

[1]*Institute of Natural Resources and Agrobiology, Spanish Council for Research, IRNAS-CSIC, Avda. Reina Mercedes 10, 41012 Sevilla, Spain*
[2]*University of Ljubljana, Faculty of Civil and Geodetic Engineering, Hajdrihova 28, 1000 Ljubljana, Slovenia*
[3]*University of Ljubljana, Biotechnical Faculty, Jamnikarjeva 101, 1000 Ljubljana*
[4]*University of Ljubljana, Faculty of Medicine, Vrazov trg 2, 1000 Ljubljana*

ABSTRACT

Trace elements play a very important role on the performance and stability of biogas digesters from a variety of biomass-containing residues, both natural or synthetic. Degradation of these complex chemical compounds occurs by the interaction of numerous microorganisms carrying out a series of pathways involving fermentative processes that ultimately lead to methane production. The purpose of this study was to provide an overview of the direct relationships existing among trace elements and enzyme activity which regulates the anaerobic digestion processes carried out by these microorganisms. Methanogenesis is one of the most trace-element enriched enzymatic pathways in biology. Trace elements are major key elements in the functioning of multiple enzymes reviewed within this work. Although exact trace-element requirements may differ slightly between pathways depending on composition and the microorganisms involved, there are some general trends characterizing the anaerobic digestion processes. Iron (Fe) is the most abundantly required metal, followed by nickel (Ni), cobalt (Co), molybdenum (Mo), tungsten (W), and zinc (Zn). In order to sustain the anaerobic digestion, trace element ions are needed for the correct structural formation and the working of those enzymes. The lack of understanding on metabolic prerequisites of microorganisms and their regulatory networks, above all at full-scale industrial anaerobic digesters, may result in consequent borderline conditions with insufficient microbial activity towards optimized methane production processes.

KEYWORDS: anaerobic digestion, enzymes, metals, methanogenesis, trace elements

3.1 INTRODUCTION

Anaerobic digestors represent a complex set of interactive actions by numerous types of microorganisms carrying out multiple metabolic pathways involved in the final production of methane, a potential source of renewable energy (Ferry, 2011). Methane and surplus heat are final products of anaerobic digestion and come with other benefits, such as a reduction in input organic loads. Also, bioremediation or biodecontamination of multiple pollutants or undesired compounds which need to be processed to reduce secondary environmental effects thus leading to an improvement in output waste which can then be used safely for other means, such as fertilizers or to meet the requirements for disposal to the environment.

The importance of anaerobic digestion is increasing due to increasing volumes of wastes that need to be processed and the diversity and variety of wastes (Verstraete *et al.*, 2002; Tabatabaei *et al.*, 2010). To achievement required levels for modern processes, anaerobic digestions need to be optimized so that the increasing amounts of wastes generated by modern society can be processed. Anaerobic digestors are major contributors to the processing of wastewaters, industrial sludges and numerous residues, including toxic and highly polluting wastes. Anaerobic digesters offer a practical solution to the processing of increasing amounts and variety of residues being generated. The same anaerobic processes take place in natural environments such as animal rumen, rice fields, peatlands, soils, and other ecosystems. Different environments, either natural or man-made, can present some differences in the microorganisms involved in the processes of degrading biomass all the way down to methane. The large microbial diversity existing in these microbial communities is well known and has been shown to be fairly stable over time in spite of moderate external or environmental changes.

Depending on the history, biogeography, and operational parameters of biogas reactors, crucial microbial constituents can fluctuate showing different abundance and level of activity as a consequence of, for instance, trace-element deficiency, long-term limitation of organic substrate or short-term overload (Murovec *et al.*, 2018; Repinc *et al.*, 2018; Sun *et al.*, 2015). This variability builds up the diversity of both microorganisms and their enzymes implicated in the processes leading to methane and the complete mineralization of organic matter (i.e. biomass).

Consequently, anaerobic digestion is a complex process involving many types of microorganisms which complement their metabolic capabilities to mineralize or degrade a huge variety of organic compounds to CO_2 and methane to the highest extent possible under anaerobic conditions within the variable environmental conditions present in anaerobic digesters (Ferry, 2011). Different microorganisms carry out different steps of the processes and this explains the existence of the numerous types of metabolisms involved in the whole anaerobic digestion process of chemically highly divergent organic materials of either plant, animal or anthropogenic origin. Obviously, different metabolic pathways are constituted by numerous enzymatic reactions. The individual enzymes participating in each of these steps of the metabolic pathways involved are the final unit pieces of the anaerobic bioprocessing. Many of these enzymes are produced by a number of functionally equivalent microorganisms and require cofactors, trace elements, activators or complements and reaction substrates that need to be available to catalyze a specific chemical reaction. Trace elements are major key elements in the functioning of multiple enzymes. Metalloenzymes are metal-dependent enzymes and a large number of metalloenzymes has been described so far. Anaerobic processes (denitrification, sulfur reduction, methanogenesis and others) contain metal-dependent enzymes (Kapoor *et al.*, 2015). Methanogenesis is one of the most trace-element enriched enzymatic pathways in biology. Although the exact trace-element requirements may differ between the methanogenic pathways, depending on the used substrates, there are some general characteristics on metal requirements. Iron (Fe) is the most abundantly required trace element, followed by nickel (Ni) and cobalt (Co), and trace amounts of molybdenum (Mo) and/or tungsten (W) and zinc (Zn). Fe is generally used for transference of electrons in Fe–S clusters (Glass & Orphan, 2012), Ni can be either bound to Fe–S clusters or in the center of porphyrin (cofactor F_{430}) that is a hallmark of methanogens. Cobalt is present in cobamides involved in methyl group transfer, whereas Zn occurs as a single structural atom in several enzymes. Molybdenum (Mo), or tungsten (W), is attached to a 'pterin' cofactor to form 'molybdopterin' or 'tungstopterin', respectively, and involved in catalyzing electron redox reactions. Other alkali metals and metalloids, such as sodium (Na) and selenium (Se), are also essential for methanogenesis (David & Alm, 2010; Dupont *et al.*, 2006, 2010; Glass & Orphan, 2012). In order to sustain metabolic activity of the cell all these ions are required.

In general, an enzyme accelerates the rate of a chemical reaction, decreasing the thermodynamic threshold for the reaction to occur. Similarly, metalloenzymes catalyze chemical reactions if the adequate trace-element atom is available in the process. Depending on the enzyme, the trace-element atom is required as a redox element, a cofactor for the enzyme, for the proper configuration of the enzyme to reach its fully functional three-dimensional structure through correct protein folding. This chapter attempts to understand the interaction of trace elements with enzymes so that these biocatalyzers achieve optimum activity and so the anaerobic digestion process develops its maximum potential.

The anaerobic process is completed by the interactive action of microorganisms each performing its characteristic metabolism. Each metabolic pathaway involves the activity of different enzymes tightly regulated to perform serially from specific substrates to final defined products. In response to environmental conditions, short-term adjustments in the anaerobic digestion process conduce to determined levels of gene expression which may induce variants of specific enzymes. Slightly different enzymes have been considered as the initial step of process heterogeneity (Delvigne *et al.*, 2014), the initiator of microbial diversity and process variability. This in turn results in ongoing redistribution of organic matter, trace elements and secondary metabolites between various microorganisms within a microbial community, generating fluctuations in the activity of the whole microbial community that ultimately results in a modified community structure over time, as reported before (Repinc *et al.*, 2018; Sun *et al.*, 2015).

This chapter deals with the importance of trace elements for the enzymes, which represents the minimum functional units involved in the process, so that the biocatalytic processes can be optimally carried out during anaerobic digestions. Microbial growth requires nutrients in the form or organic and inorganic compounds to produce energy and biomass for growth. Energy can be used to support maintenance functions by prokaryotes in the cells as well as to contribute to building the biomolecular blocks required for growth through a progressive increase in biomass and final cell division into daughter cells. The capability of microorganisms (i.e. Bacteria and Archaea) to self-maintain and to obtain and process the substances available in their surroundings are the basis for the optimization of major biological processes required in the regeneration of industrial and society residues, such as in anaerobic digesters.

3.2 MAJOR PATHWAYS AND TRACE-ELEMENT REQUIREMENTS IN ANAEROBIC DIGESTION

During anaerobic digestions of chemically complex substrates there are many microbial metabolic pathways involved in the production of methane, such as wastewater treatment plant sludge, paper mill sludge, organic fraction of municipal solid waste, biogas energetic plants, fats and oils, dairy wastewaters, spoiled food and drinks, and waste streams from food and pharmaceutical industries (Ferry, 2011). Many of the steps involved in anaerobic digestion require the availability of specific trace elements to maintain the process and at a rather high rate under the complex environmental conditions of industrial scale reactors.

In order to simplify the complexity of reactions participating in methane production and the mechanisms of trace-element involvement in anaerobic digesters, the following major pathways need to be introduced. First, the bundle of hydrolysis, acidogenesis, acetogenesis where the organic load mostly formed by complex organic molecules, that is, polymers, are degraded into smaller molecules easily taken up by most microorganisms. This represents the basic organic carbon processing routes summarized in Section 3.2.1. Second, the nitrogen cycle is of critical interest in anaerobic systems, above all because some major pathways of nitrogen processing are exclusively performed in anaerobiosis. This is the case for denitrification, dissimilatory reduction of nitrate to ammonia, and anaerobic ammonia oxidation processes, essential for the removal of large amounts of nitrates or ammonia during anaerobic digestion, without methane production (Bothe *et al.*, 2007). Third, sulfate reduction represents another central microbial metabolism in anaerobic systems, especially as it interacts with various trace elements that precipitate as sulphides and thus become unavailable to microbes due to the production of H_2S by dissimilatory sulfate-reducing bacteria (DSRB) which are common competitors of methanogenesis (Barton & Fauque, 2009; Lens & Kuenen, 2001; Muyzer & Stams, 2008). Last, the methanogenic pathway, or methanogenesis, represents the final steps for the production of methane that is carried out exclusively by methanogenic Archaea under strict anaerobic conditions.

3.2.1 Organic carbon processing

A major series of preliminary steps during anaerobic digestions is the hydrolysis of complex organic matter down to small molecules or monomers that can be taken up by microorganisms as building blocks for biosynthesis to produce biomass and turnover the biomolecules required to maintain the cellular biological machinery.

Cells incorporate trace elements using specific transporters and sensors which are often discovered as metal-resistance genes (Waldron & Robinson, 2009). Their novelty suggests that there is a broad field of research to be studied in the future years. Cells have evolved mechanisms to recover the required trace elements from the environment to meet their needs. Another point that has been demonstrated is that the removal of trace elements from a bacterial medium, for instance by metal chelation, leads to the inhibition of bacterial growth, a sign of critically reduced metabolic activity (Corbin *et al.*, 2008). Thus, bacteria require the presence of certain concentrations of trace elements for proper functioning at the cellular level within the biogas production environment. This is ultimately due to the metallic prerequisites for the correct functioning of multiple enzymes.

A large fraction of biomass supplied to anaerobic digestion is represented by residual plant biomass from crop or food industries. Thus, the residues to be processed contain a highly divergent fraction of lignin, hemicellulose and cellulose, all randomly intertwined within plant cell walls for structural and antimicrobial effects. These polymers are the most abundant components of litter and need to be depolymerized, that is, chemically broken down in order to be metabolized (Cooke & Whipps, 1993; Fioretto et al., 2005). Microorganisms experience great difficulties in degrading lignin compounds under the anaerobic conditions of biogas digesters, hence the extent of lignin derivatives in the form of aromatic compounds increases over time. The process of lignin and cellulose decomposition in biological systems has attracted attention because their hydrolysis frequently limits plant biomass degradation and consequently, the processing of wastewaters, industrial organic sludge and residues containing a large proportion of plant biomass. Usually, anaerobic reactors must perform the whole digestion of input residues down to the end products of fermentations which represent the basic substrates used for methanogenesis (Acetate, CO_2 and H_2 and one- or two-C molecules such as methylamines, methanol, formate, as some examples) (Murovec et al., 2018). The complex metabolism results in a number of metabolites (Murovec et al., 2018) produced by highly active interactions between different groups of microorganisms which center their ecological niche in maintaining degradation of specific type of chemical bonds between or within complex molecules. The sequential decomposition of major polymers involves many and diverse sets of microorganisms (Guo et al., 2015; Madigan et al., 2003). Figure 3.1 shows a simplified and generalized scheme of the processes involved in anaerobic decomposition of complex polymers (e.g. hemicellulose, xylan and cellulose) down to methane through small organic acids and substrates adequate for methanogenesis.

3.2.1.1 Complex polymers

Although most polysaccharides are decomposed by extracellular enzymes that do not show trace-element requirements, the microorganisms involved in this decomposition require basic sets of trace elements at the level of their cellular metabolism. Microorganisms require trace elements for many of their enzymes to work properly and an important set of metalloenzymes, central to the metabolism of cells, are those involved in redox processes where Fe, Ni, Co and Zn are typically required (Garuti et al., 2018).

The processes of lignin and cellulose enzymatic degradation have also been reported to be influenced by trace elements. For instance, Berg et al., (1995) reported that high concentrations of Mn were essential for the correct functioning of some lignin-degrading enzymes, such as Mn peroxidases, which are also directly involved in the oxidative decomposition of lignin. Other authors (Quinlan et al., 2011) have observed that Cu greatly enhances the oxidative degradation of cellulose by some oxidative metalloenzymes. Research has also shown that divalent trace element salts (i.e. Ca, Mg) participate in the formation of metal-lignin complexes which enhances the degradation of lignocellulosic materials (Liu *et al.*, 2010) because the trace element bound improves the accessibility of enzymes to target substrate sites, a major limiting factor for lignin biodegradation. Recent investigations have shown that supplement of Ba (10 mM) greatly enhances the activity of cellulases and esterases (Muñoz *et al.*, 2016), enzymes that complement the enzymatic decomposition of cellulose and lignin. Besides fermentations that produce low-molecular weight end products (i.e. small organic acids, ethanol, CO_2, H_2) from sugars, a major route for the production of acetate in anaerobic systems is the anaerobic oxidation of fatty acids, which is often performed by syntrophic relationships with methanogenic Archaea and sulfate-reducing bacteria depending on the existing conditions. The process and the syntrophic relationship have been reported to require Mg (0.15% w/v; Stieb & Schink, 1985).

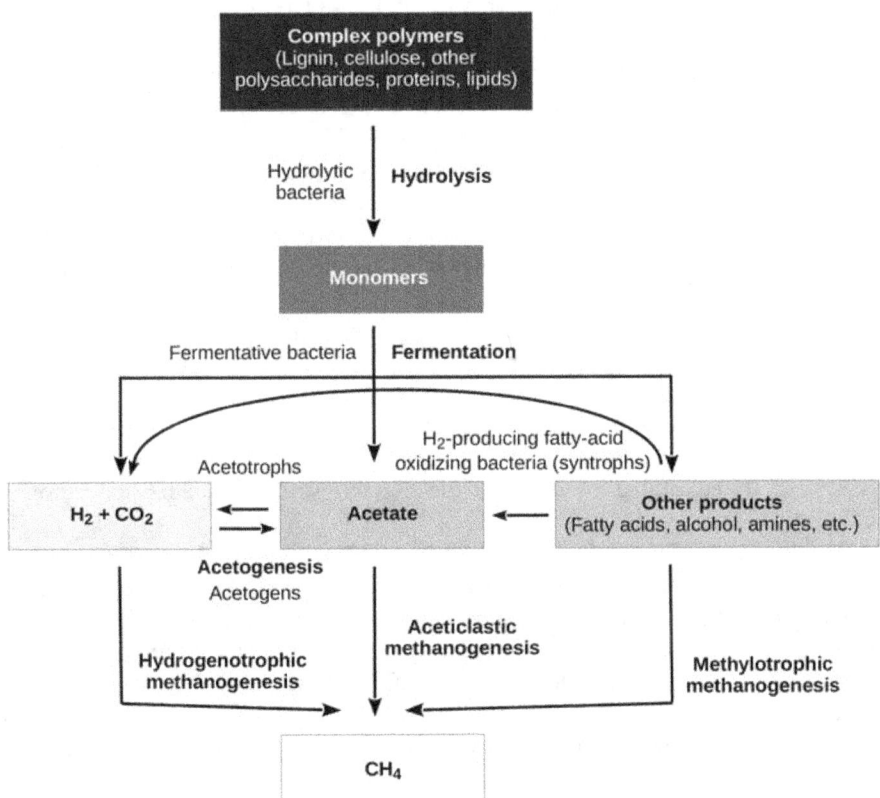

Figure 3.1 A model of serial decomposition of complex organic matter (i.e. plant biomass) to end products of fermentation and the formation of methane during anaerobic digestion.

Simultaneously to the fermentation process carried out under anaerobic conditions, and responsible for breaking down polymers into small fermentation products, the bacteria involved in these processes require the use of electron-transport systems which also require specific trace elements to maintain their function. Typically, the requirement for Fe-dependent hydrogenases and redox transformations involving cytochromes (Lehninger, 2000) during the growth of anaerobic bacteria could represent a major limitation to good performance in anaerobic bioreactors.

3.2.2 Nitrate and sulfate reduction

The sporadic inclusion of substrates that contain elevated concentration of oxidized forms of nitrogen or sulfur into anaerobic digesters results in their reduction to ammonia or hydrogen sulfides, and trace-element precipitates with a concomitant reduction in methane production. Enzymes involved in nitrate reduction to either N_2 (denitrification) or ammonia (DNRA, Dissimilatory Nitrate Reduction to Ammonium) or sulfate reduction to sulfide represent complex enzymatic pathways that all contain numerous metalloenzymes.

Enzymes involved in denitrification and DNRA contain nitrate, nitrite, nitric oxide, nitrous oxide reductases that require Mo-bis-molybdopterin guanine dinucleotide cofactor and at least one 4Fe-4S cluster, copper (CuNIR, nitrite reductase) or iron (heme cd1 NIR, nitrite reductase), iron (heme cNOR, nitric oxide reductase) and copper (NOSZ, nitrous oxide reductase) (Kielemoes *et al.*, 2000). Besides, DNRA pathway contains an additional complex set of enzymes requiring Mo, Fe, Fe-S clusters.

Enzymes accomplishing S reduction are complex structures with large molecular mass and possess at least two different polypeptides in an a2b2 tetramer containing [4Fe–4S] centers and siroheme in sulfate oxidation. The tetraheme cytochrome c3 represents the constitutive enzyme group in elemental sulfur reduction. Both groups are dependent also on the three classes of hydrogenases that contain [Fe], [NiFe] and [NiFeSe] that are essential for effective sulfate oxidation to hydrogen sulfide (Glass *et al.*, 2014).

The existence of N and S reduction pathways that rely heavily on the uptake of trace elements from the pool present in the anaerobic system represent an obstacle to simple systematic saturation of the anaerobic digestion processes by trace-element augmentation. Microorganisms performing various enzymatic reactions, thus, constantly compete for trace elements between themselves and with environmental conditions that may lead to sequestration of trace elements through either polyvalent aromatic compounds and siderophores, or precipitation with sulfur.

3.2.3 Methanogenesis

Methanogenesis is the final and most critical pathway of the anaerobic digestion because it is the process leading to the production of methane. Methanogenesis has been classified in three metabolic groups according to the use of acetate, hydrogen and CO_2 or methyl-groups (methanol, methylamines, formate, etc.), namely aceticlastic, hydrogenotrophic and methylotrophic pathways, respectively. All three pathways end in the final common steps that result in the release of methane, which is governed by the methyl coenzyme M reductase and the Zn-containing heterodisulfide reductase, both common to all methanogenic pathways. Extensive studies on the whole methanogenic pathway implicated enzymes that required novel cofactors and additional trace-element requirements. Figure 3.2 shows the enzymes characteristics of the methanogenic pathways with indications of major trace-element requirements. Trace-element requirements of the three general methanogenic pathways are relatively overlapping therefore unified and simplified observations are presented in this chapter rather than small differences between the three routes or specific differences in particular species traits.

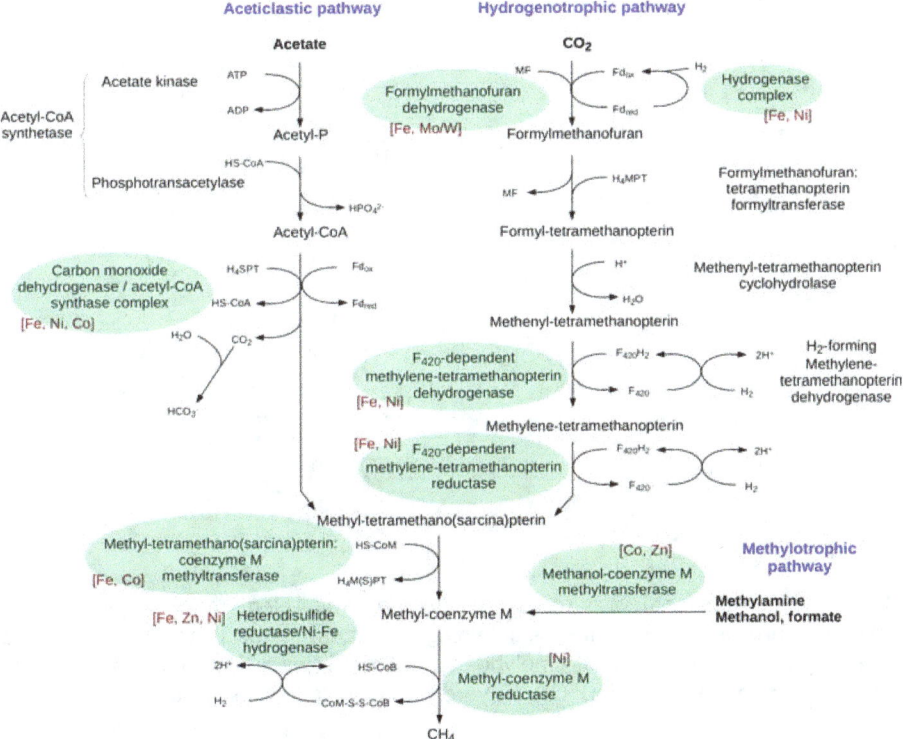

Figure 3.2 Representation of the enzymatic steps involved in the three methanogenic pathways with indication of the enzymes and their trace element requirements.

The trace element requirements of the enzymes involved in the methanogenesis have been reviewed previously (Glass & Orphan, 2012) providing the basis for the requirements of different trace elements as inhibitors or activators of the corresponding enzymes according to their depletion or supplementation, respectively.

Methanogenesis has revealed the existence of unique trace-element-requiring enzymes (Zerkle *et al.*, 2005) which has required intensive work in the field (Ferry, 2010; Thauer, 1998). Generally, Fe, Ni, Co, Mo or W and Zn are the most important trace elements required in the process (following order of abundance). Along the methanogenesis pathway, redox reactions represent some of the critical steps. The presence of Fe is typically observed forming part of the Fe-S clusters which are involved in electron transfer and are a common feature in most enzymes in the pathway. Ni binds to some Fe-S clusters and to the porphyrin ring forming the unique methanogenesis cofactor F_{430} (see below and Figure 3.3).

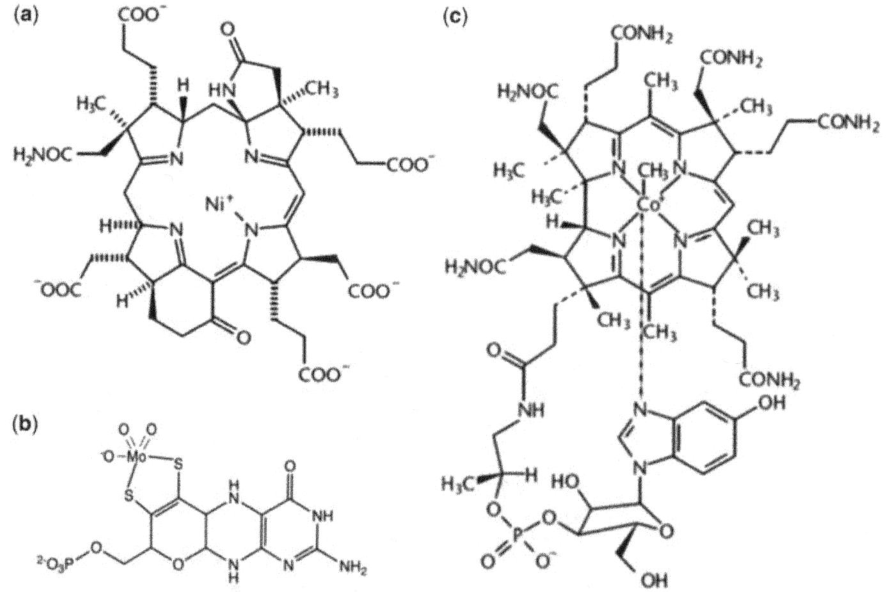

Figure 3.3 Additional examples of unique trace-element requirements in methanogenesis. Trace-element-containing cofactors unique to the methanogenic pathways: A – factor F_{430}; B – molybdenum-pterin (replace Mo with W to visualize the tungsten-pterin structure); C – 5-hydroxybenzimidazolylcobamide (Factor III).

All the methanogenic species described so far belong to the Archaea domain (Madigan *et al.*, 2003). Most methanogenic Archaea are able to perform only one of the three pathways although the exception is represented in the

Methanosarcinales; for example, *Methanosarcina acetivorans* (Galagan *et al.*, 2002) is able to carry out methanogenesis using the three described pathways metabolizing a broad range of substrates. Other unexpected results are, for instance, the report of methane production and high-density growth by *Methanocaldococcus jannaschii* from starch and organic supplements (Mukhopadhyay *et al.*, 1999), which contrasts with the typical one- or two-C molecules typically used by methanogens and suggests that further research is needed on the physiology of methanogens.

The first enzyme in the hydrogenotrophic pathway from CO_2, formylmethanofuran dehydrogenase binds with up to 9 Fe_4S_4 clusters including their links to polyferredoxin and to a Mo or W-pterin (Figure 3.3) subunit depending on the species (Vorholt *et al.*, 1996). Besides, the formyl methanofuran dehydrogenase is intimately related to four different energy-converting hydrogenases containing multiple Fe_4S_4 clusters and a Ni atom in addition to its dependency to polyferredoxins, which contain multiple (additional 6–14) Fe_4S_4 clusters forming a highly evolved electron-transfer system from H_2 (Daas *et al.*, 1994; Ferry, 1999).

Down the hydrogenotrophic pathway, a multi-aggregated hydrogenase containing Ni and Fe in its active site and four Fe_4S_4 clusters (Fox *et al.*, 1987) is in charge of reducing cofactor F_{420} with H_2. Under limitation of specific trace elements (i.e. Ni), this hydrogenase can be partially replaced by others with lower or different trace-element requirements, but some trace elements are always required in this step (Afting *et al.*, 1998; Shima *et al.*, 2008; Thauer *et al.*, 2010).

In the aceticlastic pathway, a carbon monoxide dehydrogenase/acetyl-CoA synthase complex actively participates in the process releasing a CO_2 molecule, transferring a methyl group down the pathway. This enzymatic complex contains multiple Fe_4S_4 clusters, Ni and Co.

Both the aceticlastic and hydrogenotrophic pathways arrive at a common enzymatic reaction catalyzed by the methyl-tetrahydromethanopterin-coenzyme M methyltransferase (or methly-tetrahydrosarcinapterin-coenzyme M methyltransferase in *Methanosarcina* species) that transfers a methyl group to coenzyme M. This methyltransferase contains Fe atoms and cobamine cofactors containing Co (Gartner *et al.*, 1993). In addition, the synthesis of methyl-coenzyme M needs multiple Co requiring methyltransferases in methylotrophic pathway.

A final and common enzyme for all methanogenic pathways is the methyl coenzyme M reductase which releases methane from the methyl-coenzyme M generating the coenzyme M-coenzyme B heterodisulfide. The methyl coenzyme M reductase contains two coenzyme F_{430} (Ermler *et al.*, 1997), unique to methanogens, with a Ni atom each (Figure 3.4) and the cellular level of F_{430}, depends on Ni availability (Diekert *et al.*, 1981; Lin *et al.*, 1989) which could be reflected in potential Ni-limiting growth and methane production.

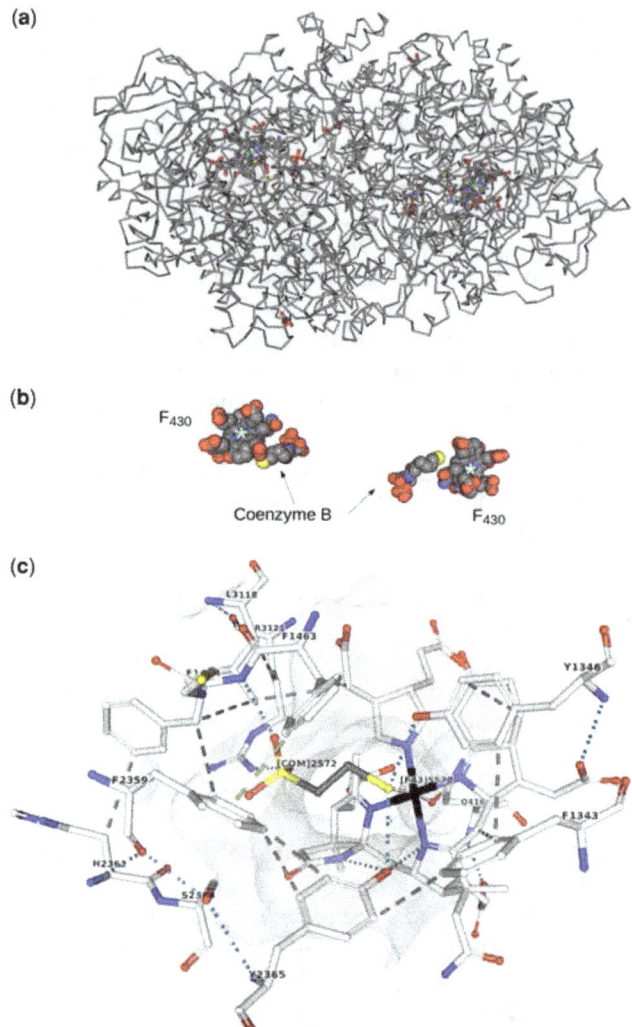

Figure 3.4 3D structure of the enzyme methyl-coenzyme M reductase from *Methanosarcina barkeri* (Accession Number 1E6Y) at 1.6 Å resolution: A – general 3D-structure of the enzyme showing the location of cofactors and trace elements; B – location of coenzyme B and coenzyme F430 within the whole enzyme structure (The Ni atoms are indicated with a white asterisk in the center of the F430 molecules); C – a single-ligand pocket view showing the interactions with the amino acidic structure ([COM] - Coenzyme M; [F43] – Factor F430 showing the central Ni atom in black; hydrogen bonds – dotted links; hydrophobic contacts – grey dashed links; trace-element interactions – white-centered dashed links). A and B are at the same scale so they can be superimposed. C is magnified to visualize interactions of amino acid residues, cofactor and trace element.

All methanogenic pathways use the heterodisulfide reductase, which contains Zn and multiple Fe_4S_4 clusters to catalyze the reduction of the heterodisulfide (i.e. coenzyme M-coenzyme B) to recycle the coenzyme M and coenzyme B molecules. This enzyme is tightly linked to a methyl-viologen hydrogenase which also contains Ni and multiple types of Fe-S clusters (Hedderich *et al.*, 2005; Thauer *et al.*, 2008) besides its relationship to additional polyferredoxins containing 12 Fe_4S_4 clusters (Reeve *et al.*, 1989). Peculiarities of subunits and their relationships to different hydrogenases exist in different methanogens and all these associated reactions sum up more trace element requirements (mainly Fe, Ni and Zn) (Glass & Orphan, 2012).

3.3 MAJOR ENZYMES INFLUENCED BY TRACE ELEMENTS

As summarized above, numerous trace-element-dependent enzymes participate in the metabolic pathways that lead to the production of methane during anaerobic digestion. In this section, the interactions taking place within some enzymes with trace element atoms are reviewed in order to present the essential participation of trace elements to maintain the enzymatic activity and structure required in the described pathways. Key enzymes of the methanogenic pathways are presented as examples of those essential interactions to understand the importance of trace elements and the mechanisms involved in trace element requirements during anaerobic digestion. An exhaustive list and evaluation of all the potential enzymes and trace elements involved in the anaerobic processing of complex organic loads is not the aim of this chapter.

3.3.1 Methyl-coenzyme M reductase

Methyl-coenzyme M reductase is a nickel tetrahydrocorphinoid-containing enzyme (coenzyme F_{430}; Figure 3.4) involved in the biological synthesis and anaerobic oxidation of methane, specifically involved in the latter, and a common step of methane production. Methyl-coenzyme M reductase catalyzes the conversion of methyl-coenzyme M and coenzyme B to methane and the heterodisulfide of coenzyme M and coenzyme B (DiMarco *et al.*, 1990; Wongnate & Ragsdale, 2015). This is the rate-limiting step and so, a main candidate to define a reduction of methane production as a result of potential scarcity of trace elements (i.e. Ni).

Structures of methyl-coenzyme M reductases have been solved and their general structures have been well described, in brief, by Wongnate & Ragsdale (2015): 'The crystal structures show that MCR is a dimer of heterotrimers ($\alpha_2\beta2\gamma_2$) with a molecular mass of 270 kDa (Ellefson & Wolfe, 1981). The three subunits ($\alpha\beta\gamma$) tightly associate to form two 50 Å hydrophobic channels (one in each heterotrimer) (Ermler *et al.*, 1997) ending in a pocket that accommodates a redox-sensitive nickel tetrapyrrole cofactor (coenzyme F_{430}), which plays an essential role in catalysis (Goubeaud et al., 1997; Becker & Ragsdale, 1998)'. The mechanisms involved in this enzyme activity have been studied (Wongnate & Ragsdale, 2015) and the enzyme requires strict anaerobic conditions to show activity. The structure envisions the complexity of this enzyme forming two symmetric active sites and showing the location of the substrates (methly-coenzyme M and coenzyme B) and the Ni-containing cofactor (coenzyme F_{430}). Figure 3.5 shows the three-dimensional structure of a methyl-coenzyme M from a methanogenic archaeon.

3.3.2 Heterodisulfide reductase

The function of this enzyme is to recycle the oxidized coenzyme M-coenzyme B compound resulting from the activity of the methyl-coenzyme M reductase to regenerate the reduced forms of coenzymes M and B. This is a required step to maintain the pathway by providing reduced forms of these coenzymes. Heterodisulfide reductase forms a complex with Ni and Fe-dependent hydrogenase, methyl-viologen hydrogenase. The visualization of these coenzyme and trace-element interactions within the heterodisulfide reductase/[NiFe]-hydrogenase tridimensional structure is shown in Figure 3.5. The high number of different types of Fe–S clusters (Fe_4S_4, Fe_3S_3, Fe_2S_2, etc.) present in the molecule and coordinated in the Cys-Cys-Gly (CCG) motifs (Wagner et al., 2017) is worthy of note. For instance, in *Methanothermobacter marburgensis*, a novel type of $[Fe4-S4]^{3+}$ was reported (Hamann et al., 2007) and the N-terminal CCG domain could be linked to a Zn site suggesting an additional interaction enzyme-metals. This is another example of the extensive requirement of trace elements in enzymes of the methanogenic pathways.

3.3.3 Formylmethanofuran dehydrogenase

The enzyme formylmethanofuran dehydrogenase catalyzes the first step of the hydrogenotrophic pathway of methanogenesis. It reduces CO_2 and contains multiple Fe_4S_4 clusters (e.g. 46 in *Methanothermobacter wolfeii*; Wagner et al., 2016) (Figure 3.6). The enzyme catalyzes the reduction of CO_2 and methanofuran to form formylmethanofuran. The crystal structure of a formylmethanofuran dehydrogenase revealed two active sites separated by a 43 Å long tunnel responsible for the transference of the formyl group. The numerous Fe_4S_4 clusters apparently couple the four tungsten redox centers present in the enzyme from *M. wolfeii* (Wagner et al., 2016) forming a spiral along the protein. The case of *Methanobacterium thermoautotrophicum* is interesting because it contains a molybdenum formylmethanofuran dehydrogenase and a tungsten formylmethanofuran dehydrogenase (Hochheimer et al., 1996).

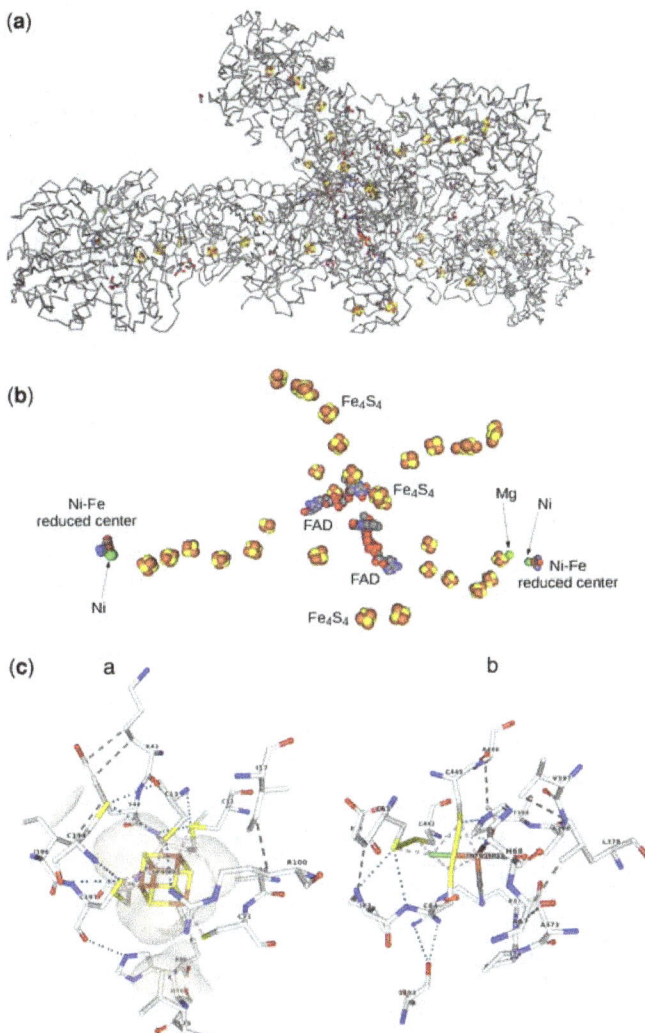

Figure 3.5 3D structure of the enzyme heterodisulfide reductase/[NiFe]-hydrogenase complex from *Methanothermococcus thermolithotrophicus* (Accession Number 5ODC) at 2.3 Å resolution: A – general 3D-structure of the enzyme; B – location of flavin (FAD), Fe4S4 clusters and Ni-Fe reduced centers within the whole enzyme structure; C – ligand pocket views showing (a) the interactions of a Fe4S4 cluster ([SF4] showed by the central cube) and (b) a Ni-Fe reduced active center ([NFU]) with the amino acidic structure. [SF4] – Fe4S4 cluster which constitutes the central cube in (C a); hydrogen bonds – dotted links; hydrophobic contacts – grey dashed links; trace-element interactions – white-centered dashed links). A and B are at the same scale so they can be superimposed. C is magnified to visualize interactions of amino acid residues and an Fe4S4 cluster.

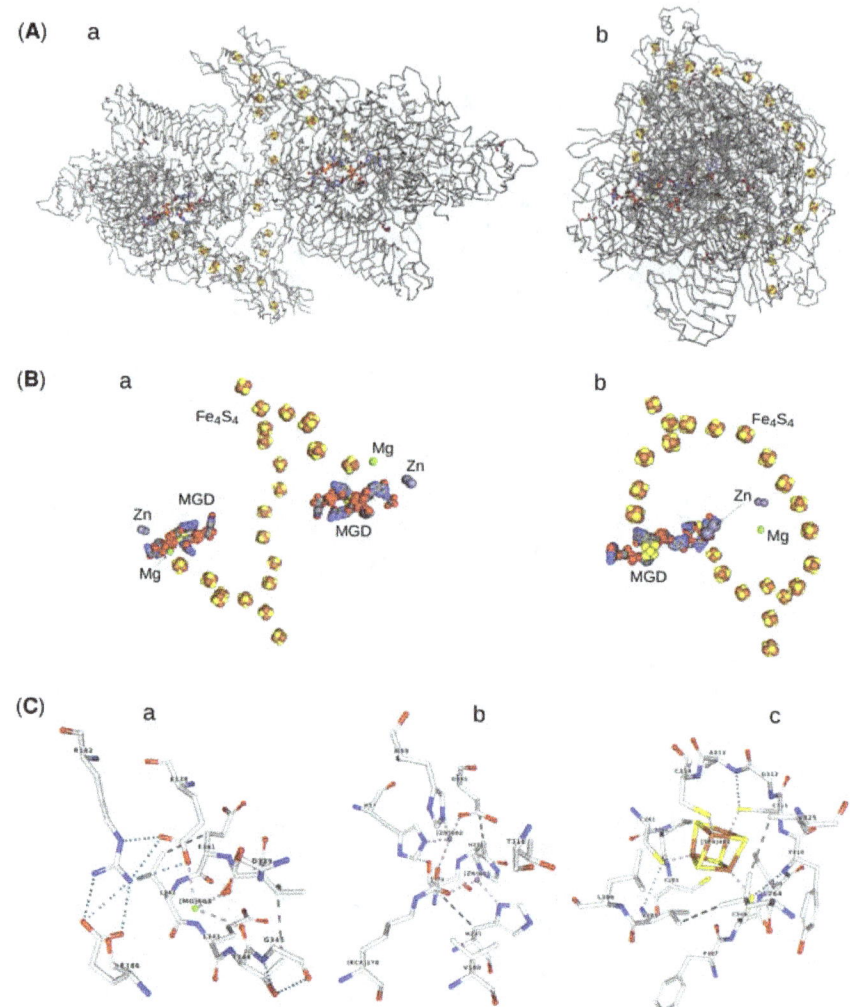

Figure 3.6 3D structure of the enzyme formylmethanofuran dehydrogenase from *Methanothermobacter wolfeii* (Accession Number 5T5I) at 1.9 Å resolution: A – general 3D-structure of the enzyme in two different perspectives (a) and (b); B – location of Fe₄S₄ clusters, Zn and Mg atoms and molybdopterin guanine dinucleotide (MGD) within the whole enzyme structure using the same perspectives that in A(a) and (b); C – ligand-pocket views showing (a) the interactions of a Mg atom (center of figure), (b) two Zn atoms (b) and (c) a Fe₄S₄ cluster (central cube) with the amino acidic structure (hydrogen bonds – dotted links; hydrophobic contacts – grey dashed links; trace-element interactions – white-centered dashed links). A and B are at the same scale so they can be superimposed. C is magnified to visualize interactions of amino acid residues and trace elements.

3.4 PERSPECTIVES

Metabolic pathways, including methanogenic pathways and all required diverse groups of microorganisms to fully mineralize biomass, represent highly coordinated networks of interactive features leading to a final product or products. This adds complexity to an already highly complex scenario of interactive behavior whose regulation is barely understood. These regulatory mechanisms remain to be studied in detail within a microorganism and between the microorganisms forming a whole natural, highly complex, community. Metabolism and its regulation is a field requiring intense research that will, eventually, drive us to a in-depth understanding on how microorganisms function and respond to their surroundings and environmental factors. The requirements of different trace elements can limit growth and responsiveness of microorganisms to their community partners, as well as reduce the functional viability of partial or full metabolic pathways. Herein, we have presented an overview of trace element–enzyme interactions which need to be comprehended in order to fully explain microbial and enzymatic response during optimization of methane-producing processes. Future research should be based on the conclusion that trace elements can (and often do) actually limit biomass degradation and methane production during anaerobic digestions.

Further research is required on trace-element interactions and enzymes both in the methanogenesis as well as in many other pathways that directly or indirectly influence methane production during anaerobic digestions of biomass (nitrogen and sulfur metabolisms). This advancement requires multidisciplinary investigations due to the multifactorial problem of the availability of trace elements in nature and complex systems (i.e. anaerobic reactors). Interactions between trace elements and other compounds (e.g. divalent cations, aromatic compounds and humic acids, microbial siderophores) lead to different levels of availability of trace elements for uptake and integration into biomass of microorganisms. Such cases force microorganisms to respond accordingly, increase their energetic expenditure and invest resources to fine-tune their protein makeup to their current environmental conditions. The lack of understanding of metabolic prerequisites of microorganisms in full-scale industrial anaerobic digesters may result in consequent borderline conditions with insufficient microbial activity towards methane production (Garuti *et al.*, 2018; Murovec *et al.*, 2018; Repinc *et al.*, 2018).

A few future options such as nanotechnology, synthetic biology and chemical engineering should be considered. Nanotechnology (Duhan *et al.*, 2017) that is used for specific delivery of particular elements or amelioration of limiting compounds contains valuable applied points to be considered. However, currently available solutions cannot be realized within the cost-benefit margins needed for implementation in actual bioreactors when applied to large-scale processes.

Synthetic biology has the potential to generate fully known microorganisms with a highly specific role in anaerobic bioreactors and is an additional alternative, perhaps in the next decade. However, this potential alternative requires precise and integrated knowledge of all processes involved and is still currently outwith our grasp.

The exploration of the fine boundary between the thresholds of methane-limiting trace element concentrations and trace element growth inhibition should be considered above all, because the chemical reactions governing the availability of trace elements in complex systems are not fully understood. The potential toxicity of some trace elements in presence of other elements or organic compounds or under specific conditions further shows the multifactorial complexity that needs to be considered when fine tuning trace-element requirements in complex systems is attempted.

Overall, our current understanding on trace-element limiting methane-production during anaerobic digestion processes is in demand of further research from several perspectives to close the existing gaps in knowledge resulting from the high complexity of anaerobic digesters. This is observed from the huge microbial diversity and variability of enzymes and metabolisms, the regulatory mechanisms involved both in controlling microbial dynamics and enzyme-encoding gene expression, as well as enzyme activity in response to the availability of the trace elements required during the whole biodegradation of biomass to ultimately generate methane.

ACKNOWLEDGEMENTS

The authors acknowledge funding from the Spanish Ministry of Science, Universities and Innovation (JMGG), the regional Government of Andalusia (BIO288)(JMGG), Slovenian Research Agency (ARRS) National project J1-6732 and P2-0180.

REFERENCES

Afting C., Hochheimer A. and Thauer R. (1998). Function of H2-forming methylenetetrahydromethanopterin dehydrogenase from *Methanobacterium thermoautotrophicum* in coenzyme F420 reduction with H2. *Archives of Microbiology*, **169**, 206–210.

Barton L. L. and Fauque G. D. (2009). Biochemistry, physiology and biotechnology of sulfate-reducing bacteria. *Advances in Applied Microbiology*, **68**, 41–98.

Becker D. F. and Ragsdale S. W. (1998). Activation of methyl-SCoM reductase to high specific activity after treatment of whole cells with sodium sulfide. *Biochemistry*, **37**, 2639–2647.

Berg B., Calvo de Anta R., Escudero A, Gärdenas A., Johansson M. B., Laskowski R., Madeira M., Mälkönen E., McClaugherty Meentemeyer V. and Virzo de Santo A. (1995). The chemical composition of newly shed needle litter of Scots pine and other pine species in a climatic transect. X. Long-term decomposition in a Scots pine forest. *Canadian Journal of Botany*, **73**, 1423–1435.

Bothe H., Newton W.E. and Ferguson S.J. (eds) (2007). Biology of Nitrogen Cycle, Elsevier, the Netherlands, 452 pp, https://doi.org/10.1016/B978-0-444-52857-5.X5000-0

Cooke R. C. and Whipps J. M. (1993). Ecophysiology of Fungi, Blackwell, Oxford, UK.

Corbin B. D., Seeley E. H., Raab A., Feldmann J., Miller M. R., Torres V. J., Anderson K. L., Dattilo B. M., Dunman P. M., Caprioli R. M., Nacken W., Chazin W. J. and Skaar E. P. (2008). Metal chelation and inhibition of bacterial growth in tissue abcesses. *Science*, **319**, 962–965.

Daas P. J. H., Hagen W. R., Keltjens J. T. and Vogels G. D. (1994). Characterization and determination of the redox properties of the 2[4Fe-4S] ferredoxin from *Methanosarcina barkeri* strain MS. *FEBS Letters*, **356**, 342–344.

David L. A. and Alm E. J. (2010). Rapid evolutionary innovation during an Archaean genetic expansion. *Nature*, **469**, 93–96.

Delvigne F., Zune Q., Lara A. R., Al-Soud W. and Sorensen S. J. (2014). Metabolic variability in bioprocessing: implications of microbial phenotypic heterogeneity. *Trends in Biotechnology*, **32**, 608–616.

Diekert G., Konheiser U., Piechulla K. and Thauer R. K. (1981). Nickel requirement and factor F430 content of methanogenic bacteria. *Journal of Bacteriology*, **148**, 459–464.

DiMarco A. A., Bobik T. A. and Wolfe R. S. (1990). Unusual coenzymes of methanogenesis. *Annual Reviews in Biochemistry*, **59**, 355–394.

Duhan J. S., Kumar R., Kumar N., Kaur P. and Nehra K. (2017). Nanotechnology: the new perspective in precision agriculture. *Biotechnology Reports*, **15**, 11–23.

Dupont C. L., Yang S., Palenik B. and Bourne P. E. (2006). Modern proteomes contain putative imprints of ancient shifts in trace metal geochemistry. *Proceedings of the National Academy of Sciences USA*, **103**, 17822.

Dupont C. L., Butcher A., Ruben R. E., Bourne P. E. and Caetano-AnollŽs G. (2010). History of biological metal utilization inferred through phylogenetic analysis of protein structures. *Proceedings of the National Academy of Sciences USA*, **107**, 10567–10572.

Ellefson W. L. and Wolfe R. S. (1981). Component C of the methylreductase system of *Methanobacterium*. *Journal of Biological Chemistry*, **256**, 4259–4262.

Ermler U., Grabarse W., Shima S., Goubeaud M. and Thauer R. K. (1997). Crystal structure of methyl-coenzyme M reductase: the key enzyme of biological methane formation. *Science*, **278**, 1457–1462.

Ferry J. G. (1999). Enzymology of one carbon metabolism in methanogenic pathways. *FEMS Microbiology Reviews*, **23**, 13–38.

Ferry J. G. (2010). The chemical biology of methanogenesis. *Planetary and Space Science*, **58**, 1775–1783.

Ferry J. G. (2011). Fundamentals of methanogenic pathways that are key to the biomethanation of complex biomass. *Current Opinion of Biotechnology*, **22**, 351–357.

Fioretto A., Nardo C. D., Papa S. and Fuggi A. (2005). Lignin and cellulose degradation and nitrogen dynamics during decomposition of three litter species in a Mediterranean ecosystem. *Soil Biology and Biochemistry*, **37**, 1083–1091.

Fox J. A., Livingston D. J., Orme-Johnson W. H. and Walsh C. T. (1987). 8-Hydroxy-5-deazaflavin-reducing hydrogenase from Methanobacterium thermoautotrophicum: 1. Purification and characterization. *Biochemistry*, **26**, 4219–4227.

Galagan J. E., Nusbaum C., Roy A., Endrizzi M. G., Macdonald P., FitzHugh W., Calvo S., Engels R., Smirnov S., Atnoor D., Brown A., Allen N., Naylor J., Stange-Thomann N., DeArellano K., Johnson R., Linton L., McEwan P., McKernan K., Talamas J., Tirrell A., Ye W., Zimmer A., Barber R. D., Cann I., Graham D. E., Grahame D. A., Guss A. M., Hedderich R., Ingram-Smith C., Kuettner H. C., Krzycki J. A., Leigh J. A., Li W., Liu J., Mukhopadhyay B., Reeve J. N., Smith K., Springer T. A., Umayam L. A., White O., White R. H., Conway de Macario E., Ferry J. G., Jarrell K. F., Jing H., Macario A. J., Paulsen I., Pritchett M., Sowers K. R., Swanson R. V., Zinder S. H., Lander E., Metcalf W. W. and Birren B. (2002). The genome of *M. acetivorans* reveals extensive metabolic and physiological diversity. *Genome Research*, **12**, 532–542.

Gartner P., Ecker A., Fischer R., Linder D., Fuchs G. and Thauer R. K. (1993). Purification and properties of N5-methyltetrahydromethanopterin: coenzyme M methyltransferase from *Methanobacterium thermoautotrophicum*. *European Journal of Biochemistry*, **213**, 537–545.

Garuti M., Langone M., Fabbri C. and Piccinini S. (2018). Methodological approach for trace elements supplementation in anaerobic digestion: experience from full-scale agricultural biogas plants. *Journal of Environmental Management*, **223**, 348–357. https://doi. org/10.1016/j.jenvman.2018.06.015

Glass J. B. and Orphan V. J. (2012). Trace metal requirements for microbial enzymes involved in the production and consumption of methane and nitrous oxide. *Frontiers in Microbiology*, **3**, 61.

Glass J. B., Yu H., Steele J. A., Dawson K. S., Sun S., Chourey K., Pan C., Hettich R. L. and Orphan V. J. (2014). Geochemical, metagenomic and metaproteomic insights into trace metal utilization by methane-oxidizing microbial consortia in sulphidic marine sediments. *Environmental Microbiology*, **16**, 1592–1611.

Goubeaud M., Schreiner G. and Thauer R. K. (1997). Purified methyl-coenzyme-M reductase is activated when the enzyme-bound coenzyme F430 is reduced to the nickel(I) oxidation state by titanium(III) citrate. *European Journal of Biochemistry*, **243**, 110–114.

Guo J., Peng Y., Ni B.-J., Han X., Fan L. and Yuan Z. (2015). Dissecting microbial community structure and methane-producing pathways of a full-scale anaerobic reactor digesting activated sludge from wastewater treatment by metagenomic sequencing. *Microbial Cell Factories*, **14**, 33.

Hamann N., Mander G. J., Shokes J. E., Scott R. A., Bennatti M. and Hedderich R. (2007). A cysteine-rich CCG domain contains a novel [4Fe-4S] cluster binding motif as deduced from studies with subunit B of heterodisulfide reductase from *Methanothermobacter marburgensis*. *Biochemistry*, **46**, 12875–12885.

Hedderich R., Hamann N. and Bennati M. (2005). Heterodisulfide reductase from methanogenic Archaea: a new catalytic role for an iron-sulfur cluster. *Biological Chemistry*, **386**, 961–970.

Hochheimer A., Linder D., Thauer R. K. and Hedderich R. (1996). The molybdenum formylmethanofuran dehydrogenase operon and the tungsten formylmethanofuran dehydrogenase operon from Methanobacterium thermoautotrophicum. Structures and transcriptional regulation. *European Journal of Biochemistry*, **242**, 156–162.

Kapoor V., Li X. M., Elk M., Chandran K., Impelliteri C. A. and Santo Domingo J. W. (2015). Impact of heavy metals on transcriptional and physiological activity of nitrifying bacteria. *Environmental Science and Technology*, **49**, 13454–13462.

Kielemoes J., de Boever P. and Verstraete W. (2000). Influence of denitrification on the corrosion of iron and stainless steel powder. *Environmental Science and Technology*, **34**, 663–671.

Lehninger A. L. (2000). Principles of Biochemistry, 3rd edn, Worth Publishers, New York.

Lens P. N. L. and Kuenen J. G. (2001). The biological sulfur cycle: novel opportunities for environmental biotechnology. *Water Science and Technology*, **44**, 57–66.

Lin D. G., Nishio N., Mazumder T. K. and Nagai S. (1989). Influence of Co2+, Ni2+ and Fe2+ on the production of tetrapyrroles by *Methanosarcina barkeri*. *Applied Microbiology and Biotechnology*, **30**, 196–200.

Liu H., Zhu J. Y. and Fu S. Y. (2010). Effect of lignin-metal complexation on enzymatic hydrolysis of cellulose. *Journal of Agricultural and Food Chemistry*, **58**, 7233–7238.

Madigan M., Martinko J. M. and Parker J. (2003). Brock Biology of Microorganisms, Prentice Hall, New Jersey.

Mukhopadhyay B., Johnson E. F. and Wolfe R. S. (1999). Reactor-scale cultivation of the hyperthermophilic methanarchaeota Methanococcus jannaschii to high cell densities. *Applied and Environmental Microbiology*, **65**, 5059–5065.

Muñoz C., Fermoso F. G., Rivas M. and Gonzalez J. M. (2016). Hydrolytic enzyme activity enhanced by Barium supplementation. *AIMS Microbiology*, **2**, 402–411.

Murovec B., Makuc D., Kolbl Repinc S., Prevoršek Z., Zavec D., Šket R., Pečnik K., Plavec J. and Stres B. (2018). 1H NMR metabolomics of microbial metabolites in the four MW agricultural biogas plant reactors: a case study of inhibition mirroring the acute rumen acidosis symptoms. *Journal of Environmental Management*, **222**, 428–435.

Muyzer G. and Stams A. J. (2008). The ecology and biotechnology of sulphate-reducing bacteria. *Nature Reviews Microbiology*, **6**, 441–454.

Quinlan R. J., Sweeney M. D., Leggio L. L., Otten H., Poulsen J-C. N., Johansen K. S., Krogh K. B. R. M., Jorgensen C. I., Tovborg M., Anthonsen A., Tryfona T., Walter C. P., Dupree P., Xu F., Davies G. J. and Walton P. H. (2011). Insights into the oxidative degradation of cellulose by a copper metalloenzyme that exploits biomass components. *Proceedings of the National Academy of Sciences USA*, **108**, 15079–15084.

Reeve J. N., Beckler G. S., Cram D. S., Hamilton P. T., Brown J. W., Krzycki J. A., Kolodziej A. F., Alex L., Orme-Johnson W. H. and Walsh C. T. (1989). A hydrogenase-linked gene in *Methanobacterium thermoautotrophicum* strain delta H encodes a polyferredoxin. *Proceedings of the National Academy of Sciences USA*, **86**, 3031–3035.

Repinc S. K., Sket R., Zavec D., Mikus K. V., Fermoso F. G. and Stres B. (2018). Full-scale agricultural biogas plant metal content and process parameters in relation to bacterial and archaeal microbial communities over 2.5 year span. *Journal of Environmental Management*, **213**, 566–574.

Shima S., Pilak O., Vogt S., Schick M., Stagni M. S., Meyer-Klaucke W., Warkentin E., Thauer R. K. and Ermler U. (2008). The crystal structure of [Fe]-hydrogenase reveals the geometry of the active site. *Science*, **321**, 572–575.

Stieb M. and Schink B. (1985). Anaerobic oxidation of fatty acids by *Clostridium bryantii sp. nov.*, a spore-forming, obligately syntrophic bacterium. *Archives of Microbiology*, **140**, 387–390.

Sun L., Pope P. B., Eijsink V. G. and Schnürer A. (2015). Characterization of microbial community structure during continuous anaerobic digestion of straw and cow manure. *Microbial Biotechnology*, **8**, 815–827.

Tabatabaei M., Rahim R. A., Abdullah N., Wright A.-D., Shirai Y., Sakai K., Sulaiman A. and Hassan M. A. (2010). Importance of the methanogenic archaea populations in anaerobic wastewater treatments. *Process Biochemistry*, **45**, 1214–1225.

Thauer R. K. (1998). Biochemistry of methanogenesis: a tribute to Marjory Stephenson. *Microbiology*, **144**, 2377–2406.

Thauer R. K., Kaster A.-K., Goenrich M., Schick M., Hiromoto T. and Shima S. (2010). Hydrogenases from methanogenic Archaea, nickel, a novel cofactor, and H2 storage. *Annual Reviews in Biochemistry*, **79**, 507–536.

Verstraete W., Doulami F., Volcke E., Tavernier M., Nollet H. and Roels J. (2002). The importance of anaerobic digestion for global environmental development. *Journal of Environmental Systems and Engineering*, **706**, 97–102.

Vorholt J. A., Vaupel M. and Thauer R. K. (1996). A polyferredoxin with eight [4Fe-4S] clusters as a subunit of molybdenum formyl-methanofuran dehydrogenase from *Methanosarcina barkeri*. *European Journal of Biochemistry*, **236**, 309–317.

Wagner T., Koch J., Ermler U. and Shima S. (2016). The methanogenic CO reducing-and-fixing enzyme is bifunctional and contains 46 [3Fe-4S] clusters. *Science*, **357**, 114–117.   [2]

Wagner T., Koch J., Ermler U. and Shima S. (2017). Methanogenic heterodisulfide reductase (HdrABC-MvhAGD) uses two non-cubane [4Fe-4S] clusters for reduction. *Science* **357**, 699–703.

Waldron K. J. and Robinson N. J. (2009). How do bacterial cells ensure that metalloproteins get the correct metal? *Nature Reviews*, **6**, 25–35.

Wongnate T. and Ragsdale S. W. (2015). The reaction mechanism of methyl-coenzyme M reductase. *Journal of Biological Chemistry*, **290**, 9322–9334.

Zerkle A. L., House C. H. and Brantley S. L. (2005). Biogeochemical signatures through time as inferred from whole microbial genomes. *American Journal of Science*, **305**, 467.

4

TE Supplementation Engineering

Jimmy Roussel[1], Cynthia Carliell-Marquet[2], Adriana F. M. Braga[3], Mirco Garuti[4], Antonio Serrano[5,6] and Fernando G. Fermoso[6]

[1]*Blue Sky Bio ltd, Chester, United Kingdom*
[2]*Severn Trent Water, United Kingdom*
[3]*Biological Processes Laboratory, Center for Research, Development and Innovation in Environmental Engineering, São Carlos School of Engineering (EESC), University of São Paulo (USP), São Carlos, Brazil*
[4]*Centro Ricerche Produzioni Animali (CRPA), Italy*
[5]*School of Civil Engineering, The University of Queensland, QLD, Australia*
[6]*Instituto de Grasa, Spanish National Research Council (CSIC), Spain*

ABSTRACT

Anaerobic digestion industries need to achieve higher performance and strive harder to play a key role in the green future of the energy sector. The importance of trace elements (TE) in the welfare of anaerobic bioreactors must be taken into account by the stakeholder/user to achieve these objectives. However, the implementation of a TE strategy is often stopped by its complexity, a lack of resource and the economic reality of a full-scale operating plant. The aim of this chapter is to support the translation of academic research findings to the engineering and operating of full-scale plant. Management tools have been developed to help operator and stakeholder in their TE assessment of their anaerobic digestion (AD) plant and suggest potential strategies to overcome deficiency. It is essential to understand the key elements of the AD system when developing the TE strategy. Feedstock is the sole natural provider of TE to the AD and determines the matrix inside the reactor. Reactor design and operating conditions fix the chemical environment that governs the behaviour and availability of the TE while controlling the TE need for bacterial population.

KEYWORDS: anaerobic bioreactor, design, full-scale, management tools, operating condition, trace element supplementation

4.1 INTRODUCTION

Energy from anaerobic digestion (AD) contributes towards the targets for renewable energy production and greenhouse gas mitigation in many countries. European countries have developed modern biogas technologies and competitive national biogas markets throughout decades of intensive research and technical development (Al Seadi *et al.* 2008). This has been achieved through support schemes, national policies, the efforts of private companies and stakeholders, and high quality research in universities/research centres.

Anaerobic digestion developed at first as a sludge-processing technology, aiming to reduce sludge volume and pathogen content prior to its disposal. The process design was simple and the biochemical knowledge minimal. In the latest four decades, anaerobic digestion has evolved into a process able to produce/recover energy through the production of methane. Research studies have been conducted to enhance the rate of biogas production, increase the speed of conversion of various feedstocks and test the ability of anaerobic digestion to treat feedstock other than sewage sludge (agricultural residue, food waste and many others). The expansion of the AD technology, through this new vector, happened throughout Europe, particularly in Germany, Italy and the United Kingdom. The number of biogas and biomethane plants at the end of 2015 were over 17,000 and 450, respectively. The generation of biogas by anaerobic digestion technology was over 18,000 million m^3 in 2015, representing 4% of the biogas share in the natural gas use (Scarlat *et al.*, 2018). The use of the biogas was divided between electricity generation (61,000 GWh) and heat production (130,000 TJ). In 2015, the European biogas sector provided 66,200 jobs representing 6% of the total jobs within the renewable energy sector; the biogas turnover in 2015 was estimated to be 6 bn € (EurObserv'ER, 2015).

Commonly, biogas plants produce electricity and heat by the combustion of biogas on-site in combined heat and power units, but anaerobic digestion can also be a source of biomethane. The purified biogas can then be injected to the gas grid (1.4 million m^3 in 2015; Scarlat *et al.*, 2018) or used as a biofuel for transportation. Biogas and biomethane are storable energy sources and they can balance the intermittent supply of other renewables such as wind and solar power (Mauky *et al.*, 2016).

Although anaerobic digestion has been adopted widely throughout the agricultural and wastewater sectors to recover bioenergy for waste, the process cost-effectiveness has considerably hindered its expansion in the energy market. AD operators often rely heavily on government subsidies in order to remain operational and compete with other sources of energy. However, government subsidies (green energy subsidies and feed in tariffs) have been reducing dramatically over the past few years, i.e. by 40% in the case of UK, jeopardizing its future growth.

If anaerobic digestion wants to play a key role in the energy of the future, it needs to become economically sustainable without any government incentives; one way to achieve this is through better and quicker degradation of the feedstock. Studies have been conducted on the physical parameters of AD to improve mixing, heating and pre-digestion treatment. AD comprises a series of sequential and interdependent microbiological reactions and it is thus necessary to ensure that the microbial community underpinning the process is as active as possible and performs optimally. Several important factors influence microbial growth and activity, including ideal conditions of pH, temperature and redox potential; carbonaceous substrates; macronutrients, such as nitrogen and phosphorus; and micronutrients i.e. trace elements (TE). The balanced availability of various nutrients coupled with the provision of ideal growth conditions is essential for anaerobic digesters. Disruptions to one or more of these factors may disturb the activity of specific groups of microorganisms and, thus, impair digester performance.

The supplementation of trace elements has a strong influence on microbial metabolism and can inhibit activity in cases of excessive concentrations or of low bioavailability. Studies showed that biogas production can be enhanced by 15–30% with effective supplementation of TE. If the energy output from the UK AD sector, as an example, could be boosted by 20%, this would generate up to 2.1TWh of additional energy. The extra generation from the same infrastructure would be able to provide green energy for 460,000 households (equivalent to a city of the size of Birmingham, Naples or Köln), generating an extra revenue of £230 million per annum. Environmental benefits will be a reduction in CO_2 emitted and the volume of biosolids for land spreading or incineration.

Engineering research needs to implement the findings from the previous chapters (TE speciation, microbial interaction) to industrial digesters. The main challenge is to scale-up the supplementation of TE on a full-scale anaerobic digester in a cost effective manner. The accurate work done in laboratory-controlled environments cannot be directly translated to a full-size anaerobic digester as the operational conditions are not completely known or can vary. The quality of the feedstock, mixing system or even the pH can vary and influence the TE requirement of the microbial community. Each digester is unique and will require a specific solution, however, using literature on previous studies and expert knowledge, the development of an efficient dosing strategy or management tool is feasible, taking into account the reactor specificity, the anaerobic digester operating conditions and the type of feedstock.

4.2 MANAGEMENT TOOLS

Process monitoring helps to understand the biochemical and physical-chemical processes occurring within a biogas plant, while helping to maintain the process stability. Thus, key parameters have been identified as early indicators of process imbalance and prediction models have been applied in practice to simulate anaerobic bioreactor performance (Drosg, 2013; Soren & Nelles, 2015). Control objectives to optimize the anaerobic fermentation are shifting from the regulation of key variables (measured on-line like temperature, mixing intensity, loading rate) to the prediction of overall process performance (generally off-line measurements like chemical analysis). Human operation is often included in the management loop when taking decisions to modify the steady state of the process. Trace elements are one of the key parameters that can influence anaerobic bioreactor performances and recent studies show the starvation of TE could be a rate-limiting step in the process (Qiang *et al.*, 2012).

The decision on whether TE supplementation of AD is required and, if it is the case, how to approach the dosing process requires the consideration of different scenarios and required outcomes. A management tool diagram has been developed covering the different scenarios typically encountered while operating large scale AD (Figure 4.1). The diagram is divided into three sections: evaluation (top), characterization (middle) and supplementation (bottom).

The first (top) area is the evaluation of general AD performance. The reasons behind an underperforming digester can be multiple. Data collection of the operating conditions, feedstock and digester parameters will allow a global assessment of the digester well-being and capabilities. TE limitation needs to be considered if no obvious causes have been found, such as temperature fluctuations, presence of inhibitor or salinity.

The second (middle) area focuses on TE requirements. Extensive elemental analysis of the feedstock is crucial to spot potential deficiency in key TEs. Analytical methodologies consider feedstock characteristics (TS, VS, BMP – biochemical methane potential) and residual methane potential to evaluate the performance of full-scale biogas plants (Ahlber-Eliasson *et al.*, 2017; Ruile *et al.*, 2015).

The third (bottom) area relates to the supplementation of TE if required. Two main directions can be taken to increase the bioavailability of identified deficient TE. The first option is to supplement a concentrated solution of TEs mixed with the feed or directly to the anaerobic digester. This is often recommended if the anaerobic digester is showing sign of struggling with potential failure. A high pulse addition of the TE cocktail might quickly recover the AD to normal performance. In a non-urgent case, the work prior to starting the TE supplementation should focus on the concentration required, the type of compounds used (salt or organic ligand), and the dosing system (continuous or pulse). The second option is to adapt the current system to increase the bioavailable section of the TE without any supplementation.

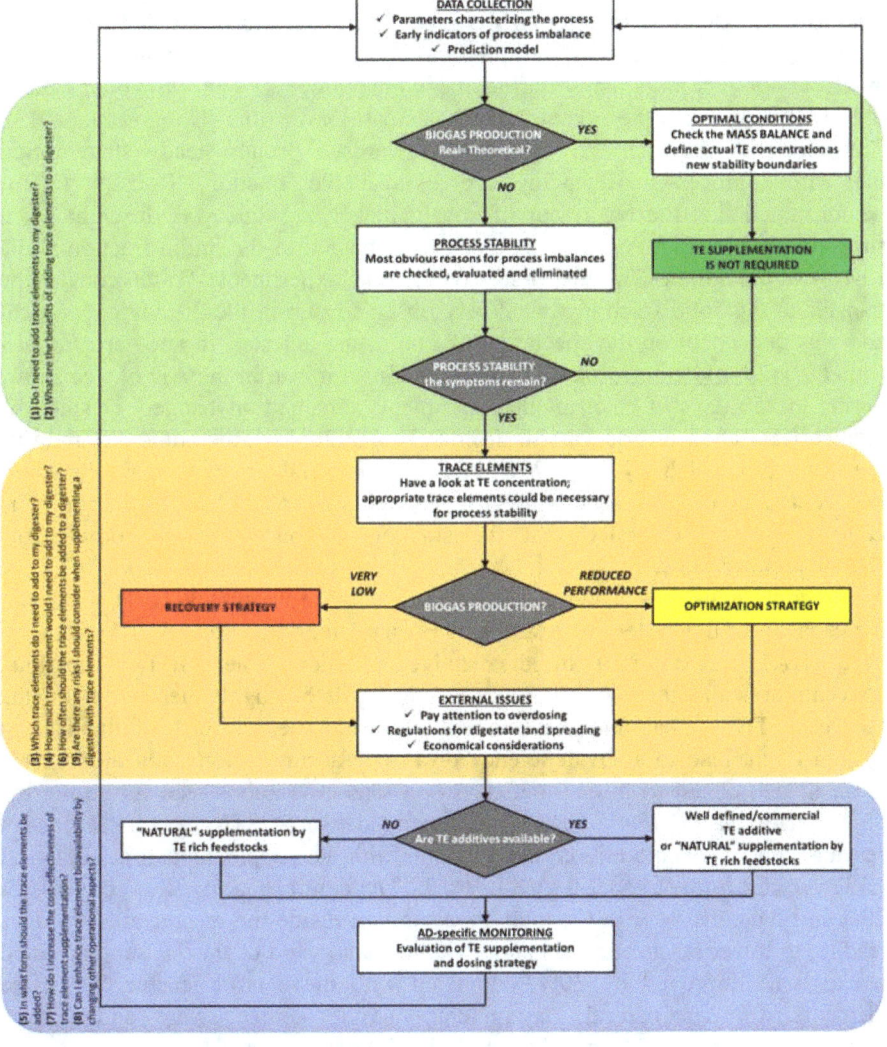

Figure 4.1 Trace-elements management tool.

4.3 INFLUENCE OF REACTOR TYPE ON TE SUPPLEMENTATION

The type of reactor plays an important role in TE supplementation as it defines the microbial community and its requirement. The approach on the best way to supplement TE will be highly dependent on the reactor used. As an example, the threshold for TE deficiency in a feedstock will vary if the system is a batch or continuous reactor.

In a batch system, the amount of TEs supplemented should be tailored to biomass/feedstock concentration and bioavailability. The total content of TE introduced in the system through the inoculum and feedstock will be identical until the end of the reaction, and the focus should be on the amount that will be available to microorganisms during the whole process. The TE/volatile solid ratio will increase as the organic matter decreases over time (Dong *et al.*, 2013). The speciation of each TE will move towards a pseudo-steady state during the whole process, influencing the bioavailable fraction. Roussel (2013) demonstrated that the behaviour of cobalt, nickel and zinc was different when supplemented as EDTA complexes. Nickel remained in the liquid fraction while zinc was precipitated in the first days of the experiments. With cobalt, the transfer to the solid fraction was slowly observed during the 30 day experiment and was dependent on the speciation of iron in the digester. In another example, Cai *et al.* (2018) demonstrated that the fractionation iron in the AD of rice straw, during the 50 days of batch mode experiment, remained unchanged. Despite the total concentration of iron varying from 9.3 to 13.2 g L^{-1}, the authors found that the iron bioavailability was low and that residual fraction was the most representative fraction, accounting for 97–98% of the total concentration. The authors also demonstrated that the supplementation of iron improved the reactor performances.

In a continuous system, the amount of TEs inside the system will be dependent on the capacity of the system to retain the TEs, particularly if the hydraulic retention time (HRT) is different to the sludge retention time (SRT). The total concentration and the bioavailability fraction will be highly dependent on the feedstock TE content and their speciation inside the reactor. An unbalanced system might lead to a TE deficiency or a toxic accumulation. González-Suárez *et al.* (2018) observed the TE deficiency in semi-continuous reactors during the digestion of maize straw. The authors operated two SCSTR reactors, a control reactor without metal addition and a reactor with TE supplementation, both with OLR varying from 0.5 to 2.0 gVS L^{-1} d^{-1}. The control reactor showed a loss of 28% in Fe and of 44% in Co total concentration inside the reactor after 96 days leading to lower reactor performance. The methane yield of the TE supplemented reactor was improved by 30%, compared with the control reactor. The trace elements, from a natural clay-mineral mixture source, were added at a concentration of 1 g L^{-1} once a week during 28 days.

In a single stage reactor, the entire full microbial consortium responsible for the AD process is subjected to the same system conditions. In perfectly mixed flow reactors, such as continuously stirred tank reactor (CSTR), the composition of the reactor is uniform and only varies over time. In a plug-flow reactor, the composition of the media varies through the reactor height (Levenspiel, 1999), thus the microorganisms located in a posterior position are subject to a concentration of TE where their bioavailability is dependent on phenomena occurring in the anterior portion of the reactor. Most full-scale reactors are in a flow pattern intermediate between perfectly mixed and plug-flow.

The sorption mechanism of TEs from the soluble to the solid phase is determinant in the retention of the TE in the anaerobic reactor. In this sense, reactors with granular sludge or support material for biomass retention allowing a higher SRT than HRT guarantee longer TE residence time in their system. The TE uptake is affected by biomass concentration in the reactor and the sorption kinetics. Long SRT should then limit the risk of washout, which is a constant challenge in a continuous system. However, a slow mass transfer of TE inside the biofilm might limit their availability to the microorganisms compared with suspended biomass.

Two UASB reactors fed with methanol were assessed by van der Veen *et al.* (2007), both supplied with TE, but one without addition of Co in the substrate and the other deprived of Ni. The authors evaluated the bioavailability of the TE according to sequential metal extraction (SME) using the Tessier modified method. The SME divides the TE into four fractions, in order of their binding strength. The author found that Co did not exhibit losses in the reactor deprived of cobalt addition; it was mostly present in a strongly bound fraction, suggesting excellent retention of this metal in the UASB reactor. On the other hand, Ni was lost relatively quickly from the UASB reactor and, correspondingly, it was extracted at 72% in the most loosely bound fraction, with only 18% in a strongly bound fraction. A submerged anaerobic membrane bioreactor (SAMBR) is more efficient in terms of biomass retention than the UASB reactors since, due the presence of the membrane, the biomass can form a biofilm, leading to higher solids retention time (SRT). Thanh *et al.* (2017) found that the TE retention during supplementation in SAMBR varied in comparison to a UASB. Zinc and iron were better retained in SAMB, while nickel and cobalt were retained to a lesser extent. The authors also demonstrated that the losses occurred when TE had a high content in loosely bound fractions.

Voelklein *et al.* (2017) carried out a single- versus two-phase system experiment to equate the TE requirement to the specific microorganisms involved in the different stages of the AD. TE supplementation was evaluated in a single- and two-stage system for digestion of food waste. Both systems failed when exceeding the OLR of 2.0 g VS L^{-1} d^{-1} and were restored after the addition of Co, Fe, Mo, Ni and Se. The specific methane yield rose to the same levels as before for both systems, with the single-phase presenting higher values. The requirements of a two-stage system, with supplementation only in the second reactor (methanogenic), was optimized to reduce the TE concentration needed since the Archaea community is more sensitive and responsible for the major stage in the AD.

4.4 INFLUENCE OF OPERATIONAL CONDITIONS ON TE SUPPLEMENTATION

Operational conditions play an important role in the quantity and quality of TE supplementation considering they affect the anaerobic microbial community dynamics (Fontana *et al.*, 2016; Xu *et al.*, 2018). In this section, the role of TEs in association with organic loading rate, temperature, pH and hydraulic retention time (HRT) are reviewed.

4.4.1 Organic loading rate

Results and conclusion from several literature studies demonstrate that the maximum organic-loading rate supported by the digester is dependent by the bioavailability of TEs. TE supplementation might be required to increase the ORL while keeping the process in balance.

Semi-continuous digestion tests were conducted in 2 L stirred tank reactors at 38°C fed with cattle slaughterhouse wastewater; increasing the OLR step-wise from 0.3 to 2.76 g COD L^{-1} d^{-1} with a corresponding decrease in HRT throughout 170 days of monitoring (Schmidt *et al.*, 2018). The TE solution consisting of Fe, Ni, Co, Mn, Mo was added daily to three digesters to reach optimal concentrations (Lemmer *et al.*, 2008) and the remaining three reactors were used as controls. The addition of TEs resulted in enhanced degradation efficiency, increased biogas production and an improved process stability. Higher OLR and lower HRT were achieved with TEs supplementation in comparison to the control digesters; control reactors exhibited a process failure at OLR 25% lower than the maximum investigated in the test and that achieved with TE supplementation (Schmidt *et al.*, 2018).

In another case, two continuously stirred tank reactors with 8 L volume fed semi-continuously at 39°C with a substrate mixture consisting of manure and industrial waste were used to evaluate the effects of TE addition in a 104 day test. The OLR was 2.4 kg VS^{-1} m^{-3} d^{-1} in the first 66 days and it was increased to 3.3 kg VS^{-1} m^{-3} d^{-1} at day 67, until the end of the test. The first reactor was used as control where only iron was supplied while the second reactor worked with iron and TEs, the composition of which was set to reach an additional 0.2 mg kg^{-1} Co and Ni concentration in the digestate (Nordell *et al.*, 2016). This experiment clearly demonstrated that the supplementation of TEs was necessary to achieve a fully stable and optimal process as the reactor supplied with TEs showing decreased volatile fatty acid (VFA) concentrations (−89%), greater biogas production rate (+24%) and better biogas yield (+10%) at upper OLR (3.3 kg VS^{-1} m^{-3} d^{-1}) if compared with the control reactor. Furthermore, the foam occurring in the control reactor at high OLR was probably due to the accumulation of long chain fatty acids or other organic intermediary, sign of an unstable process (Nordell *et al.*, 2016). The TE supplemented reactor did not suffer from foaming.

4.4.2 Temperature

Two main process temperatures are used in anaerobic fermentation and digestion systems, mesophilic conditions, 35–40°C and thermophilic conditions, 50–65°C (Bischofsberger *et al.*, 2004; Gerardi, 2003). However, the temperature range between 40°C and 50°C is also important to take into consideration because of the self-induced temperature increase sometimes observed during anaerobic digestion. This phenomenon, occurring at full-scale, can cause severe process disturbances (Lindorfer *et al.*, 2008). The digester temperature has a direct influence on the microbial community present in the reactor and so the TE requirement. Moreover, the chemical reactions determining TE speciation are also temperature dependent.

A well known advantage of thermophilic systems is higher conversion rates or higher OLR in comparison to mesophilic systems (Ahn, 2000). The higher conversion rates could be due to quicker microbial growth rates because thermophilic methanogens in general have a faster duplication rate if compared with mesophilic methanogens (Borja *et al.*, 1995). Several studies revealed an increase in degradation rates for thermophilic processes, compared with mesophilic system. This increase cannot be associated with the catalytic rate of enzymatic reactions as mesophilic and thermophilic enzymes are fully comparable at their optimum temperatures, regarding structure, catalytic route and thermo-sensitivity (Elias *et al.*, 2014; Zavodszky *et al.*, 1998). The effect on enzyme production under sub-optimal growth conditions is unknown but if enzyme production is assumed suboptimal in both cases, the amount of TEs that are needed for a thermophilic system to be optimised could, in general, be considered higher compared with similar mesophilic system (Hendriks *et al.*, 2018).

Literature suggests different nutritional requirement for various temperature ranges. Minimum requirements for iron, nickel, cobalt, and zinc in thermophilic methane fermentation from glucose has been observed to be ten times higher than those necessary for mesophilic anaerobic fermentation from acetate, indicating a possible decrease in bioavailability or increase in nutrient requirements at thermophilic temperatures (Takashima & Shimada, 2004).

In a laboratory-scale study with 5 L digesters that work at high ammonia concentrations (0.5–0.9 g NH_3 L^{-1}) with the OLR set to 2.3 g VS L^{-1} d^{-1} and an HRT of 30 days, the TEs supplementation containing Fe, Co, Ni, Se, W showed differences even within the mesophilic range. An increase in the operating temperature of only 5°C (from 37°C to 42°C) without TE addition had no impact on methane yield, but resulted in a large amount of volatile fatty acids accumulation, particularly propionate, which imbalanced the process at 42°C. The TE addition at 42°C resulted in 12–23% increased methane yield and an improvement of the degree of degradation with a decrease in propionate concentration from 4 g L^{-1} to below 0.1 g L^{-1} (Westerholm et al., 2015).

In full-scale anaerobic digesters the individual temperature depends on plant specific parameters such as feedstock type and quantity, self-heating, insulation potential and availability of cooling system. The seasonal variation of the external temperature and other factor complicates the maintenance of the optimal temperature (Lindorfer et al., 2008) and those fluctuations affect the process stability due to the temperature sensitivity of methanogenic Archea (Speece, 1996). TEs supplementation could be a reasonable strategy to reduce the negative impacts of operating temperature changes.

4.4.3 Two phase anaerobic digestion/pH role

The physical separation of hydrolytic/acid-producing and methane-producing microrganisms in two different reactors, where optimum environment conditions for each group of organisms would be provided, is considered an interesting technology for overall process improvement. The hydrolytic/acidogenic reactor is often operated under moderately acidic conditions (pH 5–6) at a short HRT (<5 days) (Pohland & Ghosh, 1971; Ponsà et al., 2008).

Changes in pH and HRT can not only affect the performance of the anaerobic process, it can also radically affect the metal speciation of TMs with interrelated effects (Thanh et al., 2016). Generally, literature studies showed that increase in pH led to the transformation of TEs from available mobile fractions to stable organic forms reducing their bioavailability. In contrast, a reduction in pH results in the dissolution of TEs into the liquid medium, shifting them towards more bioavailable fractions (Dong et al., 2013; Lopes et al., 2008). Taking in consideration other parameters, the redox reaction between Fe and S lead to iron sulphide (FeS) formation and the interaction between precipitated FeS and TEs resulting in the formation of insoluble FeS–metal complexes (Dong et al., 2013) with limited bioavailability for hydrolytic bacteria.

Recent studies investigated the determination of TEs in the first phase of anaerobic digestion, where hydrolysis and acidification are considered together due to their close connection (Burgess & Pletschke, 2008; Frey & Hegeman, 2007): TEs involved in hydrolytic enzyme reaction mechanism include Mg, Zn, Ni, Co, Se, Mo, and Fe.

A substrate consisting in maize silage (75%), grass cuttings (15%), winery by-products (5%), cow manure (2.5%), and other bio-wastes (2.5%) was used in anaerobic digestion experiments with batch reactors to evaluate the influences of TEs concentration and composition on the hydrolysis rate. Results showed that at high volatile fatty acids concentration (200 mmol/L) the increase in Co concentration is required to ensure good hydrolysis of the substrate and its conversion to VFA. Moreover, the simultaneously increasing of Co and Se concentrations will improve substrate hydrolysis and acidification rate (Ezebuiro & Körner, 2017). This study concluded that the original composition and concentration of TEs in substrates could be unbalanced and lower than the required level for continuative substrate hydrolysis; different TEs combination addition, i.e. Co and Se, could be beneficial for the process.

4.4.4 Hydraulic retention time

Studies on the relationship between HRT and TEs supplementation in anaerobic processes are very limited. The evaluation of Cu and Zn distribution in swine wastewater in an up-flow biodigester at different HRTs reported a decreased metal-retention capacity of the bioreactors when the HRT was reduced from 18 days to 7.5 day (Cestonaro do Amaral et al., 2014). This finding is important for the operation of biogas plants in two-phase anaerobic digestion systems: the acidogenic reactors with short HRT could need a higher concentration of TEs to compensate for the fraction of the metals which become less bioavailable (Thanh et al., 2016), primarily when the hydrolysis rate represents the limiting factor of the process.

The removal of VFAs from the acidogenic reactor to feed the second stage methanogenic reactor is an important parameter in a two-phase system to control VFAs concentration and pH in the first stage, as well as the OLR in the methanogenic reactor (Cysneiros et al, 2012). In full-scale digesters, the partial recirculation of the effluent from the methanogenic stage to the acidic stage can help to buffer the rapidly produced VFAs, maintain a suitable pH and enhance the process to overcome the bacterial wash out problem (Zuo et al., 2013). At the same time the recirculation can guarantee a partial recycling of TEs for the system, limiting the lack of nutrients that occur in the acidogenic reactor with short HRT. A proper management of periodicity and volume of recirculation with its nutrient recycling leads to the creation of favourable conditions for microbial growth and nutrient transport for both acidogenic and methanogenic phases.

Many operational parameters have been confirmed to be associated directly or indirectly with anaerobic microbial evolution and methane generation. They are often set to drive the fermentation towards one direction rather than another, sometimes with the purpose of recovering an unbalanced system, or in other cases to optimize the process. These alterations in many cases bring unpredictable physico-chemical variations of the environment and shifts in microbial population that affect the nutrient requirement of the digester. The metal–biomass interaction and sulphide chemistry trigger reactions of precipitation, co-precipitation, adsorption and uptake that influence the bioavailable fraction of TEs which can be utilized in the anaerobic digestion process. At full scale, the awareness that some nutrients are retained in the digester and others will be released with the effluent digestate according to applied operational conditions, and that the metals could be more or less bioavailable depending on their speciation has to be clear. Improvement in the knowledge of these mechanisms will lead to tailor-made TE formulations and dosing strategies developed to have a specific concentration of TEs in their bioavailable form to best fit each anaerobic digestion process with its operational conditions.

4.5 IMPORTANCE OF FEEDSTOCK CHARACTERIZATION ON TE SUPPLEMENTATION

Trace elements are usually supplied with feedstock to the biogas plants, being necessary for its evaluation to ensure a stable and optimal anaerobic digestion process (Schattauer et al., 2011). The metabolic effect of the trace elements is dependent on the concentration of trace elements in the feedstock, but also on the availability and speciation of the trace elements (Demirel & Scherer, 2011). Complexation reactions (in the liquid phase or solid phase) play an important role in bioreactors making a particular trace element either more or less bioavailable (Fermoso et al., 2015). Other main mechanisms involved in the bioavailability are the chelation of metals, the ion exchange, the adsorption, the inorganic micro-precipitation and the translocation of trace elements into the microorganisms (Zandvoort et al., 2006).

The shortfall supplementation of trace elements can results in the destabilization of the process due to their necessary role as cofactors in different metabolic pathways (Fermoso et al., 2015; Matheri et al., 2016). An excess of trace elements in the supplementation could entail their accumulation up to inhibitory concentrations (Fermoso et al., 2015). Another concern of an excessive supplementation would be an undesirable accumulation of trace elements in the final digestate, which could limit land application (Serrano et al., 2014). Moreover, an excess of trace-element supplementation is costly. Therefore, it is very important that the correct use of the management tools are used, as described in section 4.3. A detailed evaluation and characterization of the anaerobic system can optimize the supplementation of the trace metals, avoiding the previously described undesirable effects.

As the feedstock is the primary source of trace elements, and many other compounds that could influence the chemical bioavailability of the trace elements, it is important to carry out a deep characterization of the feedstock. Although each feedstock has its own character, four different types of feedstock can be defined that will influence trace-element bioavailability in digestion: sulphur-rich, phosphate-rich, lipid-rich and lignocellulosic waste.

4.5.1 Sulphide-rich feedstock

Sulphur is an essential macronutrient for the growth of methanogenic microorganisms (Mountfort & Asher, 1979). However, the presence of sulphur compounds in the reactor could affect the bioavailability of trace elements, even at very low concentrations (Gonzalez-Gil *et al.*, 2003; Zandvoort *et al.*, 2006). Most trace elements predominantly react with sulphide to form insoluble salts and sulphide concentration is a determining factor in the TE speciation and bioavailability (Fermoso *et al.*, 2009; Roussel & Carliell-Marquet, 2016). Typical sulphide-rich feedstocks are pig slurry, algae (fucoidan content) and acid-mining drainage and examples of TE concentrations are shown Table 4.1.

One of the main factors affecting trace-elements bioavailability is the presence of sulphate-reducing bacteria (SRB), which reduce the sulphate to sulphide in anaerobic conditions (Chen *et al.*, 2008; Dall'Agnol & Moura, 2014). The precipitation of trace elements in the form of metal sulphides enhances the retention of the trace elements in the digesters, although these trace elements may be no longer bioavailable for the uptake by microorganisms (Zandvoort *et al.*, 2008). Studies demonstrated that presence/addition of sulphide reduced the bioavailability of several TEs (such as zinc, iron, copper) by a translocation from the carbonate to the sulphide fraction (Zandvoort *et al.*, 2006, 2008). Therefore, the monitoring of the trace elements concentration based on the total concentration could not be representative enough of the requirements of the microorganisms, therefore the monitoring of the soluble concentration or, if possible, a more detailed fractionation study is recommended (Pinto-Ibieta *et al.*, 2016).

4.5.2 Phosphate-rich feedstock

Phosphorous is a basic nutrient for the growth of the anaerobic microorganisms and it also plays a key role in the immobilisation of the biomass on mineral particles (Chen *et al.*, 2008). The anaerobic digestion of phosphate-rich feedstock can result in high concentrations of soluble phosphate in the digesters (Carliell-Marquet & Wheatley, 2002). Sewage sludge is the main phosphate-rich feedstock due to chemical phosphorus removal processes occurring in wastewater treatment plants. The concentration of phosphate in the sludge can be increased 10-fold or more (Roussel & Carliell-Marquet, 2016). Fishery waste can also have a high phosphate content and examples of these feedstock are shown in Table 4.2. The increase in the concentration of phosphate has a direct effect on the concentration and bioavailability of others cations, such as calcium, magnesium and manganese, and hence, of the trace elements (Carliell-Marquet & Wheatley, 2002; Marti *et al.*, 2008; Muhmood *et al.*, 2018).

Roussel and Carliell-Marquet (2016) demonstrated that iron speciation was influenced by the presence of high phosphate concentration with the formation of vivianite, especially in sewage sludge. The presence of iron as vivianite increased its mobility of iron in the sludge in comparison with iron-sulphide salts. This shift on the iron speciation might change other TEs bioavailability through co-precipitation or chemical reactions (Roussel, 2013).

Table 4.1 Trace-elements composition of different sulphide-rich feedstocks.

		S ppt	N ppt	P ppt	K ppt	Al ppt	Cu ppm	Zn ppm	Mg ppm	Fe ppm	Mn ppm	Co ppb	Ni ppb
Marcato et al. (2008)	Pig slurry	8.1	2.8	29.5	37.4	0.868	590	1500	14,200	2500	0.629		
Radis Steinmetz et al. (2009)	Pig slurry				0.925	0.146	34.8	43.9	17.4	171	386	251	244
Formentini et al. (2017)	Pig slurry		3.7	0.6	1.7		21.0	43.0					
Ometto et al. (2018)	Algae (S. latissima)	2.66		0.46	7.7	130	6.02	11.9	1680	476	3.22	<420	840
Barbot et al. (2015)*	Laminaria japonica Waste	8.9	20.0	2.1	89.9		5	28	6800	3440	150	1500	3000
Migliore et al. (2012)	Algae	1.82	1.74				9.8	88		3920			4.3
Li et al. (2014)	Synthetic saline wastewater	0.6			1.46		0.024	0.024	0.009	0.014	0.042	10.8	22.6
Ueki et al. (1988)	Acid main drainage	157					550	9100		9.704	13.8		
Smyntek et al. (2018)	Acid main drainage	222	<1	<1		37.4				19,800	2.6		
Fernández-González et al. (2018)	Acid mining effluent	2.7	0.5	<0.001			5.23	20.91			8.53		14,204
Moraes et al. (2015)	Sugar beet vinasse	1.76	26.4	0.13	11.2	0.21	0.41		0.05	15.32	9.75	<0.5	660
Garcia-Depraect et al. (2017)	Vinasse	0.286	0.188	0.341	0.371		<1.0	2.3	136.2	14.3	<1.0	<1.0	<1.0

*Dry basis.

Table 4.2 Trace-element composition of different phosphate-rich feedstock.

	S ppt	N ppt	P ppt	K ppt	Al ppt	Cu ppm	Zn ppm	Mg ppm	Fe ppm	Mn ppm	Co ppb	Ni ppb
González et al. (2017) Sewage sludge		33.4	97.8	3.81		261	456					19,000
Serrano et al. (2014) Sewage sludge		7.94	4.23			95	167					64,000
Martí et al. (2010) Sewage sludge		0.104	0.071	0.072				60				
Wan et al. (2011) Thickened sewage sludge		4.0	1.231	0.32		17.8	54.5	291	628	54.3	<0.2	5680
Muhmood et al. (2018) Poultry slurry		4.5–5	0.22	2.58		1.69	2.91	0.014				550
Wang et al. (2016) Acido-genic effluent	0.06	0.7	1.246	0.121	1.53	1.00	0.28	90.45	12.39	<0.10		<200
Bohutskyi et al. (2015) Waste-water media	0.029	0.023	0.006			0.064	0.035	11	1.0	0.19	1.8	
Alvarenga et al. (2017) Primary sludge		0.14	0.07	0.02	354	0.448	1.416	11.8	44.25	0.885		
Goddek et al. (2016) Aqua-culture waste-water	0.199	0.057	0.192	1.44				396				
Vivekanand et al. (2018) Fish ensilage	4.90	64	9.2	7.95	135	11.0	0.098	1200	420	10	60	510
Madariaga & Marín (2017) Fish farming sludge		9.62	7.02			3.77	140					
Andreji et al. (2006) Fish muscle						0.128	1.01		1.64	0.056	24	14
Nges et al. (2012) Fish waste	1.8	23.8	7.7	1.6		0.9	88	380				60

Magnesium is one of the most affected cations by the high phosphate concentration due to its precipitation in the form of struvite and/or other salts (Muhmood *et al.*, 2018; Wang *et al.*, 2016). The precipitation of struvite (magnesium ammonium phosphate hexahydrate) has been widely proposed as a phosphorous recovery and valorisation technique from phosphate-rich feedstock given its potential as fertilizer (Yilmazel & Demirer, 2013). Other trace elements can precipitate associated to the precipitation of the struvite. For example, Muhmood *et al.* (2018) reported the removal of 40, 45, 66, 30 and 20% of zinc, copper, lead, chrome or nickel, respectively, during the precipitation of struvite in a reactor treating poultry slurry. Similar results were also reported by Liu *et al.* (2011) at the struvite recovery from swine wastewater. The co-precipitation of the trace elements and joint extraction with the struvite could result in a deficit of trace elements in the long-term operation of the anaerobic processes. Even if the phosphate is not recovered in the form of struvite, high concentrations of phosphate have been related to a strong decrease in the bioavailability of magnesium and manganese (Carliell-Marquet & Wheatley, 2002). Therefore, trace-element supplementation for anaerobic digesters treating phosphate-rich feedstock should take in account expected high precipitation rates, which could result in higher supplementation costs.

4.5.3 Lipid-rich feedstock

The substrates with a high concentration of lipids are usually selected as feedstock for anaerobic digestion due to their high biodegradability and methanogenic potential (Cirne *et al.*, 2007). However, the rapid and easy biodegradability of a lipid-rich feedstock can result in the destabilization of the anaerobic process due to the accumulation of short volatile fatty acids (Chan *et al.*, 2018; Cirne *et al.*, 2007). For this, anaerobic digesters at full scale are sometimes operated at low organic loading rates or with poor concentrations of lipid-rich feedstock (Karlsson *et al.*, 2012). Food industry waste is often lipid-rich feedstock, especially if the process is related to oil or dairy. Example of TE concentration in these feedstocks are shown in the Table 4.3.

The supplementation with trace elements has been studied as a strategy to ensure stable operation and to maximize methane production when lipid-rich feedstocks are treated (Chan *et al.*, 2018; Karlsson *et al.*, 2012). Iron, cobalt and nickel have been described as having high relevance in the degradation of volatile fatty acids due to their role as cofactors in different acetogenesis and methanogenesis enzymes (Karlsson *et al.*, 2012; Pinto-Ibieta *et al.*, 2016). For example, these three trace elements act as a cofactor of the carbon monoxide dehydrogenase, which it is involved in the degradation of acetate in acetogens and methanogens (Karlsson *et al.*, 2012; Ortner *et al.*, 2015). However, other authors have reported that the addition of nickel and cobalt was not enough to prevent the accumulation of fatty acids, supplementation with a more complex

Table 4.3 Trace-element composition of different lipid-rich feedstock.

		S ppt	N ppt	P ppt	K ppt	Al ppt	Cu ppm	Zn ppm	Mg ppm	Fe ppm	Mn ppm	Co ppb	Ni ppb
Vivekanand et al. (2018)	Whey	1.00	4.0	8.3	30.5	55.5	8.2	1.75	1600	1950	10	300	2750
Güler and Sanal (2009)	Whey (cow)		0.4	0.68	1.75		0.1	6.57	343	0.27	0.17	130	
	Whey (sheep)		1.2	1.19	1.97		0.17	7.84	477	0.30	0.11	160	
	Whey (goat)		1.1	0.69	2.23		0.13	4.26	334	0.24	0.11	130	
Eftaxias et al. (2018)	Emulsified slaughter-house wastes		9.02				0.135	2.30		0.430	0.127	5	8
Zhang et al. (2011)	Food waste	0.59	6.4	1.49	0.546	4.31	3.06	8.27	62.5	3.17	0.96	<30	190
Zhu et al. (2008)	Food waste				1.112		1.19	2.53	87.7	9.47	0.84		
Zhang et al. (2007)	Food waste	2.5	9.7	1.6	2.8	1202	31	76		766	60		2000
Wan et al. (2011)	Fat, oil, and grease		1.0	0.03	0.06		579	2498	13.3	29.6	<0.2	<0.2	559
Schmidt et al. (2018)	Slaughter-house wastewater							0.49		4.5	0.28	1.89	7.44
Zhang et al. (2008)	Palm oil mill effluent	0.40	0.87	0.27	5.53	6299	5.08	6.83	1.065	61.17	8572		
Pinto-Ibieta et al. (2016)	Olive mill solid waste	0.39	5.7	0.66			3.82	5.67	1687	224	3.99	16.9	858
Gowda et al. (2004)	Coconut cake			1.3			26	56	0.35	516			
Christodoulou et al. (2014)	Sunflower cake	0	7.99	3.21	4.87	147				8179			

trace element solution being more effective (Schattauer *et al.*, 2011). Zinc, molybdenum and selenium have also been reported as essential trace elements for the correct anaerobic digestion of slaughterhouse waste (Ortner *et al.*, 2015). In fact, the addition of molybdenum, cobalt and selenium have been successfully used to revert high volatile fatty acid accumulation in anaerobic digesters (Garuti *et al.*, 2018). The supplementation of trace elements for these substrates, especially iron, cobalt and nickel, should compensate for the deficiencies in feedstock composition that could result in the accumulation of volatile fatty acids.

4.5.4 Lignocellulosic waste

Vegetable-waste material is composed of different biomass generated from agricultural and agro-industrial activities. This kind of feedstock has been widely used as anaerobic feedstock, even at full scale (Garuti *et al.*, 2018). However, the mono-digestion of vegetable waste can be a challenge due to its high fibre content, i.e. cellulose, hemicellulose and lignin, as well as a deficiency in trace elements (Cai *et al.*, 2017). The unavailability of trace elements in biogas digesters can be one of the main reason for poor process efficiency in the biomethanization of lignocellulosic material (Demirel & Scherer, 2011). The low concentration of trace elements in many lignocellulosic wastes (Table 4.4) makes necessary their supplementation to maintain biogas production and to avoid destabilization of the process (Demirel & Scherer, 2011; Garuti *et al.*, 2018).

Biomethanization of lignocelullosic waste entails all the stages of the anaerobic digestion, where many enzymes and cofactors are involved requiring different trace elements in each stage. Nickel, cobalt, iron, molybdenum and/or selenium have been reported to enhance biogas production and operational performance in an anaerobic digester treating different green-waste materials. For example, the anaerobic digestion of Napier grass produced 40% more methane after a daily addition of a solution containing nickel, cobalt, molybdenum and selenium (Demirel & Scherer, 2011; Wilkie *et al.*, 1986). However, more recent studies reported that the addition of nickel and cobalt has a limited effect on the anaerobic digestion of lignocellulosic material (Cai *et al.*, 2017). These trace elements are related with the acetogenesis and methanogenesis stages, whereas the hydrolysis has been defined as the rate-limiting step for this type of feedstock (Hendriks & Zeeman, 2009; Ortega *et al.*, 2008). For lignocellulosic waste, supplementation of iron enhanced the biogas production and the biodegradability of the substrate. For example, Khatri *et al.* (2015) reported an improvement in methane production by up to 32% for the anaerobic digestion of maize straw with the addition of 1000 mg/L of Fe. However, Cai *et al.* (2017) reported that the addition of 5 mg Fe/L was enough to improve the anaerobic digestion of rice straw up to 176%. The huge differences in the reported supplementation concentration highlight the necessity of determining the bioavailability of the

Table 4.4 Trace-element composition of different substrates in lignocellulosic waste.

		S ppt	N ppt	P ppt	K ppt	Al ppt	Cu ppm	Zn ppm	Mg ppm	Fe ppm	Mn ppm	Co ppb	Ni ppb
Moraes et al. (2015)	Straw	0.76		0.44	7.9	53.8	1.95		580	50.02	11.9	<0.5	430
Serrano et al. (2014)	Strawberry waste		2.30	1.24			3.6	<1					11,180
Vivekanand et al. (2018)	Cellulose	0.03	1.0	0.0	0.01	2.25	0.0	0.0	0.0	0.0	0.0		0.0
Nges et al. (2012)	Jerusalem artichoke	0.28	5.34	0.38	1.43		1.24	4.8	1480			10	230
Wall et al. (2014)	Grass silage						5.7	23.5		277	38.4	<250	3600
Wandera et al. (2018)	Maize straw		1.5		11.1		7.3	69.7	2316	3372	71	0	54.8
Xi et al. (2015)	Wheat straw		4.9				0	65.51	0.83	7470	49.02	0	4.05
Cai et al. (2017)	Rice straw		6.41				0.55	25		294	156	0	5280
Mancini et al. (2018)	Rice straw		10.4				15.5	57.6		443		<1.0	2000
Mussoline et al. (2014)	Rice straw		5.0	0.91	8.15		15.4	45.8	1325	269		164	3709
Janke et al. (2015)	Sugarcane straw	1.61	3.98	0.46	5.00		8.2	7.7	1140	11,460	136	2201	5499
Janke et al. (2015)	Sugarcane filter cake	0.05	5.08	0.17	740		12.7	38.1	971	7880	163	971	4133
Li et al. (2018)	Grass	3.89	7.94		7.9		12.2	182	1720	168	213	7400	24,500
Wahid et al. (2018)	Grass mixture	4.72	4.5	3.81	36.4	274	11.10	32.00	2060	444	112	<300	2740

trace elements, instead of only the total concentration. For the agricultural feedstock, it could be also interesting to determine the fibre composition, since the preponderance of a concrete lignocellulosic compound could require the supplementation of different trace elements. Lignin and cellulose-rich feedstock should be supplemented with trace elements involved in different metalloenzyme-catalyzing hydrolytic reactions, such as manganese (Dismukes, 1996; Romero-Güiza et al., 2016). On the another hand, substrates with a high content of hemicellulose, which can be easily solubilized, should be supplemented with trace elements involved in the acidogenesis and methanogenesis stages (Temudo et al., 2009).

4.6 CONCLUSION

The consideration on the role of TE in a full-scale anaerobic digestion plant is still minimal due to a lack of awareness or knowledge. This chapter highlights key points AD operators and stakeholders should focus on when considering TE in their system. Understanding the complete TE speciation in anaerobic digesters requires time, equipment and expertise, but a simple analysis of operating conditions and feedstock should be enough to give an indication on potential inhibition or deficiency.

The quantification of TEs entering a system (through feedstsock analysis) and the retention potential of the process (reactor type) should allow for estimation of TE quantity in the anaerobic bioreactor. The standard operating condition of the system should evaluate the TE requirement for the microbial community and potential deficiency. Research studies are now developing computer models to afford a better picture of TE speciation/bioavailability based on input parameters (see chapter 5) and give operators an easier platform to assess their own system. Validation and distribution of these models to the AD industry will be a great leap forward in process optimisation as it has been with the publishing of the AD1 model.

Despite great progress in the understanding of TE behaviour and bioavailability, there remains a need to raise awareness of the crucial importance of TE balance in a system. It is the duty of research scientist to be able to cooperate and communicate efficiently with the industry about potential problems and, if possible, offer adequate solutions.

REFERENCES

Ahlber-Eliasson K., Nadeau E., Leven L. and Schnurer A. (2017). Production efficiency of Swedish farm-scale biogas plants. Biomass and Bioenergy, **97**, 27–37.

Ahn J. (2000). A comparison of mesophilic and thermophilic anaerobic upflow filters. Bioresource Technology, **73**, 201–205. http://dx.doi.org/10.1016/S0960-8524(99) 00177-7.

Al Seadi T., Rutz D., Prassl H., Köttner M., Finsterwalder T., Volk S. and Janssen R. (2008). Biogas Handbook, University of Southern Denmark, Esbjerg, Denmark.

Alvarenga E., Øgaard A. F. and Vråle L. (2017). Effect of anaerobic digestion and liming on plant availability of phosphorus in iron- and aluminium-precipitated sewage sludge from primary wastewater treatment plants. *Water Science and Technology*, **75**(7), 1743–1752.

Andreji J., Stranai I., Kacániová M., Massányi P. and Valent M. (2006). Heavy metals content and microbiological quality of carp (Cyprinus carpio, L.) muscle from two southwestern Slovak fish farms. *Journal of Environmental Science and Health – Part A Toxic/Hazardous Substances and Environmental Engineering*, **41**(6), 1071–1088.

Barbot Y. N., Thomsen C., Thomsen L. and Benz R. (2015). Anaerobic digestion of laminaria japonica waste from industrial production residues in laboratory- and pilot-scale. *Marine Drugs*, **13**(9), 5947–5975.

Bischofsberger W., Dichtl N., Rosenwinkel K. H. and Seyfried C. F. (2004). Anaerobtechnik, Springer, Berlin.

Bohutskyi P., Liu K., Nasr L. K., Byers N., Rosenberg J. N., Oyler G. A., Betenbaugh M. J. and Bouwer E. J. (2015). Bioprospecting of microalgae for integrated biomass production and phytoremediation of unsterilized wastewater and anaerobic digestion centrate. *Applied Microbiology and Biotechnology*, **99**(14), 6139–6154.

Borja R., Martín A., Banks C. J., Alonso V. and Chica A. (1995). A kinetic study of anaerobic digestion of olive mill wastewater at mesophilic and thermophilic temperatures. *Environmental Pollution*, **88**, 13–18. http://dx.doi.org/10.1016/0269-7491(95)91043-K

Burgess J. E. and Pletschke B. I. (2008). Hydrolytic enzymes in sewage sludge treatment: A mini-review. *Water SA*, **34**, 343–350. http://eprints.ru.ac.za/1371/.

Cai Y., Hua B., Gao L., Hu Y., Yuan X., Cui Z., Zhu W. and Wang X. (2017). Effects of adding trace elements on rice straw anaerobic mono-digestion: Focus on changes in microbial communities using high-throughput sequencing. *Bioresource Technology*, **239**, 454–463.

Cai Y., Zhao X., Zhao Y., Wang H., Yuan X., Zhu W., Cui Z. and Wang X. (2018). Optimization of Fe^{2+} supplement in anaerobic digestion accounting for the Fe-bioavailability. *Bioresource Technology*, **250**, 163–170. http://dx.doi.org/10.1016/j.biortech.2017.07.151

Carliell-Marquet C. M. and Wheatley A. D. (2002). Measuring metal and phosphorus speciation in P-rich anaerobic digesters. *Water Science and Technology*, **45**(10), 305–312.

Cestonaro do Amaral A., Kunz A., Radis Steinmetz R. L. and Justi K. C. (2014). Zinc and copper distribution in swine wastewater treated by anaerobic digestion. *Journal of Environmental Management*, **141**, 132–137.

Chan P. C., de Toledo R. A. and Shim H. (2018). Anaerobic co-digestion of food waste and domestic wastewater – Effect of intermittent feeding on short and long chain fatty acids accumulation. *Renewable Energy*, **124**, 129–135.

Chen Y., Cheng J. J. and Creamer K. S. (2008). Inhibition of anaerobic digestion process: A review. *Bioresource Technology*, **99**(10), 4044–4064.

Christodoulou C., Grimekis D., Panopoulos K. D., Vamvuka D., Karellas S. and Kakaras E. (2014). Circulating fluidized bed gasification tests of seed cakes residues after oil extraction and comparison with wood. *Fuel*, **132**, 71–81.

Cirne D. G., Paloumet X., Björnsson L., Alves M. M. and Mattiasson B. (2007). Anaerobic digestion of lipid-rich waste—Effects of lipid concentration. *Renewable Energy*, **32**(6), 965–975.

Cysneiros D., Banks C. J., Heaven S. and Karatzas K. A. G. (2012). The effect of pH control and 'hydraulic flush' on hydrolysis and Volatile Fatty Acids (VFA) production and profile in anaerobic leach bed reactors digesting a high solids content substrate. *Bioresource Technology*, **123**, 263–271, http://dx.doi.org/10.1016/j.biortech.2012.06.060

Dall'Agnol L. T. and Moura J. J. G. (2014). Sulphate-reducing bacteria (SRB) and biocorrosion. In: Understanding Biocorrosion, T. Liengen, D. Féron, R. Basséguy and I. B. Beech (eds), Woodhead Publishing, Oxford, pp. 77–106.

Demirel B. and Scherer P. (2011). Trace element requirements of agricultural biogas digesters during biological conversion of renewable biomass to methane. *Biomass and Bioenergy*, **35**(3), 992–998.

Dong B., Liu X., Dai L. and Dai X. (2013). Changes of heavy metal speciation during high solid anaerobic digestion of sewage sludge. *Bioresource Technology*, **131**, 152–158.

Dismukes G. C. (1996). Manganese Enzymes with Binuclear Active Sites. *Chemical Reviews*, **96**(7), 2909–2926.

Drosg B. (2013). Process monitoring in biogas plant. *Technical Brochure*. IEA Bioenergy, pp. 1–38, ISBN 978-1-910154-03-8.

Eftaxias A., Diamantis V. and Aivasidis A. (2018). Anaerobic digestion of thermal pre-treated emulsified slaughterhouse wastes (TESW): Effect of trace element limitation on process efficiency and sludge metabolic properties. *Waste Management*, **76**, 357–363.

Elias M., Wieczorek G., Rosenne S. and Tawfik D. S. (2014). The universality of enzymatic rate–temperature dependency. *Trends Biochemical Science*, **39**, 1–7, http://dx.doi.org/10.1016/j.tibs.2013.11.001.

EurObserv'ER (2015). The state of renewable energies in Europe. 15th EurObserv'Er report, Paris, France.

Ezebuiro N. C. and Körner I. (2017). Characterisation of anaerobic digestion substrates regarding trace elements and determination of the influence of trace elements on the hydrolysis and acidification phases during the methanisation of a maize silage-based feedstock. *Journal of Environmental Chemical Engineering*, **5**, 341–351, http://dx.doi.org/10.1016/j.jece.2016.11.032

Fermoso F. G., Bartacek J., Jansen S. and Lens P. N. L. (2009). Metal supplementation to UASB bioreactors: From cell-metal interactions to full-scale application. *Science of the Total Environment*, **407**(12), 3652–3667.

Fermoso F. G., van Hullebusch E. D., Guibaud G., Collins G., Svensson B. H., Carliell-Marquet C. M., Vink J. P. M., Esposito G. and Frunzo L. (2015). Fate of trace metals in anaerobic digestion. *Advances in Biochemical Engineering/Biotechnology*, **151**, 171–195.

Fernández-González R., Martín-Lara M. A., Iáñez-Rodríguez I. and Calero M. (2018). Removal of heavy metals from acid mining effluents by hydrolyzed olive cake. *Bioresource Technology*, **268**, 169–175.

Fontana A., Patrone V., Puglisi E., Morelli L., Bassi D., Garuti M., Rossi L. and Cappa F. (2016). Effects of geographic area, feedstock, temperature, and operating time on microbial communities of six full-scale biogas plants. *Bioresource Technology*, **218**, 980–990, http://dx.doi.org/10.1016/j.biortech.2016.07.058

Formentini T. A., Legros S., Fernandes C. V. S., Pinheiro A., Le Bars M., Levard C., Mallmann F. J. K., da Veiga M. and Doelsch E. (2017). Radical change of Zn speciation in pig slurry amended soil: Key role of nano-sized sulfide particles. *Environmental Pollution*, **222**, 495–503.

Frey P. A. and Hegeman A. D. (2007). Enzyme Reaction Mechanism, Oxford University Press, London.

García-Depraect O., Gómez-Romero J., León-Becerril E. and López-López A. (2017). A novel biohydrogen production process: Co-digestion of vinasse and Nejayote as complex raw substrates using a robust inoculum. *International Journal of Hydrogen Energy*, **42**(9), 5820–5831.

Garuti M., Langone M., Fabbri C. and Piccinini S. (2018). Methodological approach for trace elements supplementation in anaerobic digestion: Experience from full-scale agricultural biogas plants. *Journal of Environmental Management*, **223**, 348–357.

Gerardi M. H. (2003). The Microbiology of Anaerobic Digesters, John Wiley & Sons, New York.

Goddek S., Schmautz Z., Scott B., Delaide B., Keesman K. J., Wuertz S. and Junge R. (2016). The effect of anaerobic and aerobic fish sludge supernatant on hydroponic lettuce. *Agronomy*, **6**(2), 37–49.

González I., Serrano A., García-Olmo J., Gutiérrez M. C., Chica A. F. and Martín M. Á. (2017). Assessment of the treatment, production and characteristics of WWTP sludge in Andalusia by multivariate analysis. *Process Safety and Environmental Protection*, **109**, 609–620.

Gonzalez-Gil G., Jansen S., Zandvoort M. H. and van Leeuwen H. P. (2003). Effect of yeast extract on speciation and bioavailability of nickel and cobalt in anaerobic bioreactors. *Biotechnology and Bioengineering*, **82**(2), 134–142.

González-Suárez A., Pereda-Reyes I., Oliva-Merencio D., Suárez-Quiñones T., da Silva A. J. and Zaiat M. (2018). Bioavailability and dosing strategies of mineral in anaerobic mono-digestion of maize straw. *Engineering in Life Sciences*, **18**(8), 562–569, doi: 10.1002/ elsc.201700018

Gowda N. K. S., Ramana J. V., Prasad C. S. and Singh K. (2004). Micronutrient Content of Certain Tropical Conventional and Unconventional Feed Resources of Southern India. *Tropical Animal Health and Production*, **36**(1), 77–94.

Güler Z. and Şanal H. (2009). The essential mineral concentration of Torba yoghurts and their wheys compared with yoghurt made with cows', ewes' and goats' milks. *International Journal of Food Sciences and Nutrition*, **60**(2), 153–164.

Hendriks A. T. W. M. and Zeeman G. (2009). Pretreatments to enhance the digestibility of lignocellulosic biomass. *Bioresource Technology*, **100**(1), 10–18.

Hendriks A. T. W. M., van Lier J. B. and de Kreuk M. K. (2018). Growth media in anaerobic fermentative processes: The underestimated potential of thermophilic fermentation and anaerobic digestion. *Biotechnology Advances*, **36**(1), 1–13, https://doi.org/10.1016/j.biotechadv.2017.08.004.

Janke L., Leite A., Nikolausz M., Schmidt T., Liebetrau J., Nelles M. and Stinner W. (2015). Biogas Production from Sugarcane Waste: Assessment on Kinetic Challenges for Process Designing. *International Journal of Molecular Sciences*, **16**(9), 20685–20703.

Karlsson A., Einarsson P., Schnürer A., Sundberg C., Ejlertsson J. and Svensson B. H. (2012). Impact of trace element addition on degradation efficiency of volatile fatty acids, oleic pretreatment and Fe dosing on batch anaerobic digestion of maize straw. *Applied Energy*, **158**(Supplement C), 55–64.

Lemmer A. D., Mathies E. D., Mayrhuber E. D., Oechsner H. D., Preissler D. and Ramhold D.

(2008). Verfahren zur Biogaserzeugung. Process for the production of biogas. Google Patents.

Levenspiel O. (1999). Chemical Reaction Engineering, 3rd edn. John Wiley & Sons, New York, doi: 10.1016/0009-2509(64)85017-X

Li J., Yu L., Yu D., Wang D., Zhang P. and Ji Z. (2014). Performance and granulation in an upflow anaerobic sludge blanket (UASB) reactor treating saline sulfate wastewater. *Biodegradation*, **25**(1), 127–136.

Li W., Lu C., An G., Zhang Y. and Tong Y. W. (2018). Integration of high-solid digestion and gasification to dispose horticultural waste and chicken manure. *Chinese Journal of Chemical Engineering*, **26**(5), 1145–1151.

Lindorfer H., Waltenberger R., Kollner K., Braun R. and Kirchmayr R. (2008). New data on temperature optimum and temperature changes in energy crop digesters. *Bioresource Technology*, **99**, 7011–7019.

Liu Y., Kwag J.-H., Kim J.-H. and Ra C. (2011). Recovery of nitrogen and phosphorus by struvite crystallization from swine wastewater. *Desalination*, **277**(1), 364–369.

Lopes S. I. C., Capela M. I., van Hullebusch E. D., van der Veen A. and Lens P. N. L. (2008). Influence of low pH (6, 5 and 4) on nutrient dynamics and characteristics of acidifying sulfate reducing granular sludge. *Process Biochemical*, **43**, 1227–1238.

Madariaga S. T. and Marín S. L. (2017). Sanitary and environmental conditions of aquaculture sludge. *Aquaculture Research*, **48**(4), 1744–1750.

Mancini G., Papirio S., Riccardelli G., Lens P. N. L. and Esposito G. (2018). Trace elements dosing and alkaline pretreatment in the anaerobic digestion of rice straw. *Bioresource Technology*, **247**, 897–903.

Marcato C. E., Pinelli E., Pouech P., Winterton P. and Guiresse M. (2008). Particle size and metal distributions in anaerobically digested pig slurry. *Bioresource Technology*, **99**(7), 2340–2348.

Marti N., Bouzas A., Seco A. and Ferrer J. (2008). Struvite precipitation assessment in anaerobic digestion processes. *Chemical Engineering Journal*, **141**(1), 67–74.

Martí N., Pastor L., Bouzas A., Ferrer J. and Seco A. (2010). Phosphorus recovery by struvite crystallization in WWTPs: Influence of the sludge treatment line operation. *Water Research*, **44**(7), 2371–2379.

Matheri A. N., Belaid M., Seodigeng T. and Ngila J.C. (2016). The Role of Trace Elements on Anaerobic Codigestion in Biogas Production. Paper presented at the Proceedings of the World Congress on Engineering, London, UK.

Mauky E., Weinrich S., Nagele H. J., Jacobi H. F., Liebetrau J. and Nelles M. (2016). Model predictive control for demand-driven biogas production in Full Scale. *Chemical Engineering & Technology*, **39**, 652–664.

Migliore G., Alisi C., Sprocati A. R., Massi E., Ciccoli R., Lenzi M., Wang A. and Cremisini C. (2012). Anaerobic digestion of macroalgal biomass and sediments sourced from the Orbetello lagoon, Italy. *Biomass and Bioenergy*, **42**, 69–77.

Moraes B. S., Triolo J. M., Lecona V. P., Zaiat M. and Sommer S. G. (2015). Biogas production within the bioethanol production chain: Use of co-substrates for anaerobic digestion of sugar beet vinasse. *Bioresource Technology*, **190**, 227–234.

Mountfort D. O. and Asher R. A. (1979). Effect of inorganic sulfide on the growth and metabolism of Methanosarcina barkeri strain DM. *Applied and Environmental Microbiology*, **37**(4), 670.

Muhmood A., Wu S., Lu J., Ajmal Z., Luo H. and Dong R. (2018). Nutrient recovery from anaerobically digested chicken slurry via struvite: Performance optimization and interactions with heavy metals and pathogens. *Science of The Total Environment*, **635**, 1–9.

Mussoline W., Esposito G., Lens P. N. L., Garuti G. and Giordano A. (2014). Electrical energy production and operational strategies from a farm-scale anaerobic batch reactor loaded with rice straw and piggery wastewater. *Renewable Energy*, **62**, 399–406.

Nges I. A., Mbatia B. and Björnsson L. (2012). Improved utilization of fish waste by anaerobic digestion following omega-3 fatty acids extraction. *Journal of Environmental Management*, **110**, 159–165.

Nordell E., Nilsson B., Nilsson Påledal S., Karisalmi K. and Moestedt J. (2016). Co-digestion of manure and industrial waste – The effects of trace element addition. *Waste Management*, **47**, 21–27, doi: 10.1016/j.wasman.2015.02.032

Ometto F., Berg A., Björn A., Safaric L., Svensson B. H., Karlsson A. and Ejlertsson J. (2018). Inclusion of Saccharina latissima in conventional anaerobic digestion systems. *Environmental Technology*, **39**(5), 628–639.

Ortega L., Husser C., Barrington S. and Guiot S. R. (2008). Evaluating limiting steps of anaerobic degradation of food waste based on methane production tests. *Water Science and Technology*, **57**, 419–422.

Ortner M., Rameder M., Rachbauer L., Bochmann G. and Fuchs W. (2015). Bioavailability of essential trace elements and their impact on anaerobic digestion of slaughterhouse waste. *Biochemical Engineering Journal*, **99**, 107–113.

Pinto-Ibieta F., Serrano A., Jeison D., Borja R. and Fermoso F. G. (2016). Effect of cobalt supplementation and fractionation on the biological response in the biomethanization of Olive Mill Solid Waste. *Bioresource Technology*, **211**, 58–64.

Pohland F. G. and Ghosh S. (1971). Development in anaerobic stabilization of organic wastes – the two phase concept. *Environment Letters*, **1**, 255–266.

Ponsà S., Ferrer I., Vázquez F. and Font X. (2008). Optimization of the hydrolytic–acidogenic anaerobic digestion stage of sewage sludge: Influence of pH and solid content. *Water Research*, **42**, 3972–3980.

Qiang L., Zhang X. L., Jun Z., Zhao A. H., Chen S. P., Lui F., Tai J., Liu J. Y. and Qian G. R. (2012). Effect of carbonate on anaerobic acidogenesis and fermentative hydrogen production from glucose using leachate as supplementary culture under alkaline conditions. *Bioresource Technology*, **113**, 37–43.

Radis Steinmetz R. L., Kunz A., Dressler V. L., de Moraes Flores É. M. and Figueiredo Martins A. (2009). Study of metal distribution in raw and screened swine manure. *CLEAN – Soil, Air, Water*, **37**(3), 239–244.

Romero-Güiza M. S., Vila J., Mata-Alvarez J., Chimenos J. M. and Astals S. (2016). The role of additives on anaerobic digestion: A review. *Renewable and Sustainable Energy Reviews*, **58**, 1486–1499.

Roussel J. (2013). Metals behaviour in anaerobic sludge digesters supplemented with trace metals to enhance biogas production. PhD thesis, School of Civil Engineering, University of Birmingham, Birmingham, UK.

Roussel J. and Carliell-Marquet C. (2016). Significance of Vivianite Precipitation on the Mobility of Iron in Anaerobically Digested Sludge. *Frontier Environment Science*, **4**, 60, doi: 10.3389/fenvs.2016.00060

Ruile S., Schmitz S., Monch-Tegeder M. and Oechsner H. (2015). Degradation efficiency of agricultural biogas plant – A full-scale study. *Bioresource Technology*, **178**, 341–349.

Scarlat N., Dallemand J.-F. and Fahl F. (2018). Biogas: Developments and perspectives in Europe. *Renewable Energy*, **129**, 457–472.

Schattauer A., Abdoun E., Weiland P., Plöchl M. and Heiermann M. (2011). Abundance of trace elements in demonstration biogas plants. *Biosystems Engineering*, **108**(1), 57–65.

Schmidt T., McCabe B. K., Harris P. W. and Lee S. (2018). Effect of trace element addition and increasing organic loading rates on the anaerobic digestion of cattle slaughterhouse wastewater. *Bioresource Technology*, **264**, 51–57.

Serrano A., Siles J. A., Chica A. F. and Martín M. A. (2014). Anaerobic co-digestion of sewage sludge and strawberry extrudate under mesophilic conditions. *Environmental Technology*, **35**(23), 2920–2927.

Smyntek P. M., Chastel J., Peer R. A. M., Anthony E., McCloskey J., Bach E., Wagner R. C., Bandstra J. Z. and Strosnider W. H. J. (2018). Assessment of sulphate and iron reduction rates during reactor start-up for passive anaerobic co-treatment of acid mine drainage and sewage. *Geochemistry: Exploration, Environment, Analysis*, **18**(1), 76–84.

Soren W. and Nelles M. (2015). Critical comparison of different model structures for the applied simulation of the anaerobic digestion of agricultural energy crops. *Bioresource Technology*, **178**, 306–312.

Speece R. E. (1996). Anaerobic Biotechnology for Industrial Wastewater, Tennessee Archae Press, Nashville.

Takashima M. and Shimada K. (2004). Minimum requirements for trace metals (Fe, Ni, Co, and Zn) in thermophilic methane fermentation from glucose. Proceedings of 10th World Congress Anaerobic Digestion, Montreal, pp. 1590–1593.

Temudo M. F., Mato T., Kleerebezem R. and van Loosdrecht M. C. M. (2009). Xylose anaerobic conversion by open-mixed cultures. *Applied Microbiology and Biotechnology*, **82**(2), 231–239.

Thanh P. M., Ketheesan B., Yanb Z. and Stuckey D. C. (2016). Trace metal speciation and bioavailability in anaerobic digestion: A review. *Biotechnology Advances*, **34**, 122–136, http://dx.doi.org/10.1016/j.biotechadv.2015.12.006

Thanh P. M., Ketheesan B., Zhou Y. and Stuckey D. C. (2017). Effect of operating conditions on speciation and bioavailability of trace metals in submerged anaerobic membrane bioreactors. *Bioresource Technology*, **243**, 810–819, doi: 10.1016/j.biortech.2017.07.040

Ueki K., Kotaka K., Itoh K. and Ueki A. (1988). Potential availability of anaerobic treatment with digester slurry of animal waste for the reclamation of acid mine water containing sulfate and heavy metals. *Journal of Fermentation Technology*, **66**(1), 43–50.

van der Veen A., Fermoso F. G. and Lens P. N. L. (2007). Bonding Form Analysis of Metals and Sulfur Fractionation in Methanol-Grown Anaerobic Granular Sludge. *Engineering Life Science*, **7**, 480–489, doi: 10.1002/elsc.200720208

Vivekanand V., Mulat D. G., Eijsink V. G. H. and Horn S. J. (2018). Synergistic effects of anaerobic co-digestion of whey, manure and fish ensilage. *Bioresource Technology*, **249**, 35–41.

Voelklein M. A., O' Shea R., Jacob A. and Murphy J. D. (2017). Role of trace elements in single and two-stage digestion of food waste at high organic loading rates. *Energy*, **121**, 185–192, doi: 10.1016/j.energy.2017.01.009

Wahid R., Feng L., Cong W.-F., Ward A. J., Møller H. B. and Eriksen J. (2018). Anaerobic mono-digestion of lucerne, grass and forbs – Influence of species and cutting frequency. *Biomass and Bioenergy*, **109**, 199–208.

Wall D. M., Allen E., Straccialini B., O'Kiely P. and Murphy J. D. (2014). The effect of trace element addition to mono-digestion of grass silage at high organic loading rates. *Bioresource Technology*, **172**, 349–355.

Wan C., Zhou Q., Fu G. and Li Y. (2011). Semi-continuous anaerobic co-digestion of thickened waste activated sludge and fat, oil and grease. *Waste Management*, **31**(8), 1752–1758.

Wandera S. M., Qiao W., Algapani D. E., Bi S., Yin D., Qi X., Liu Y., Dach J. and Dong R. (2018). Searching for possibilities to improve the performance of full-scale agricultural biogas plants. *Renewable Energy*, **116**, 720–727.

Wang S., Hawkins G. L., Kiepper B. H. and Das K. C. (2016). Struvite precipitation as a means of recovering nutrients and mitigating ammonia toxicity in a two-stage anaerobic digester treating protein-rich feedstocks. *Molecules*, **21**(8), 1011.

Westerholm M., Müller B., Isaksson S. and Schnürer A. (2015). Trace element and temperature effects on microbial communities and links to biogas digester performance at high ammonia levels. *Biotechnology Biofuels*, **8**, 154, doi: 10.1186/s13068-015-0328-6

Wilkie A., Goto M., Bordeaux F. M. and Smith P. H. (1986). Enhancement of anaerobic methanogenesis from napiergrass by addition of micronutrients. *Biomass*, **11**(2), 135–146.

Xi Y., Chang Z., Ye X., Du J., Chen G. and Xu Y. (2015). Enhanced methane production from anaerobic co-digestion of wheat straw and herbal-extraction process residues. *BioResources*, **10**(4), 7985–7997.

Xu R., Yang Z. H., Zheng Y., Liu J. B., Xiong W. P., Zhang Y. R., Lu Y., Xue W. J. and Fan C. Z. (2018). Organic loading rate and hydraulic retention time shape distinct ecological networks of anaerobic digestion related microbiome. *Bioresource Technology*, **262**, 184–193. https://doi.org/10.1016/j.biortech.2018.04.083

Yilmazel Y. D. and Demirer G. N. (2013). Nitrogen and phosphorus recovery from anaerobic co-digestion residues of poultry manure and maize silage via struvite precipitation. *Waste Management & Research*, **31**(8), 792–804.

Zandvoort M. H., van Hullebusch E. D., Fermoso F. G. and Lens P. N. L. (2006). Trace metals in anaerobic granular sludge reactors: Bioavailability and dosing strategies. *Engineering in Life Sciences*, **6**(3), 293–301.

Zandvoort M. H., van Hullebusch E. D., Gieteling J., Lettinga G. and Lens P. N. L. (2008). Effect of Sulfur Source on the Performance and Metal Retention of Methanol-Fed UASB Reactors. *Biotechnology Progress*, **21**(3), 839–850.

Zavodszky P., Kardos J., Svingor Á. and Petsko G. A. (1998). Adjustment of conformational flexibility is a key event in the thermal adaptation of proteins. *Proceedings of the National Academy of Sciences*, **95**, 7406–7411.

Zhang R., El-Mashad H. M., Hartman K., Wang F., Liu G., Choate C. and Gamble P. (2007). Characterization of food waste as feedstock for anaerobic digestion. *Bioresource Technology*, **98**(4), 929–935.

Zhang Y., Yan L., Chi L., Long X., Mei Z. and Zhang Z. (2008). Startup and operation of anaerobic EGSB reactor treating palm oil mill effluent. *Journal of Environmental Sciences*, **20**(6), 658–663.

Zhang L., Lee Y. W. and Jahng D. (2011). Anaerobic co-digestion of food waste and piggery wastewater: Focusing on the role of trace elements. *Bioresource Technology*, **102**(8), 5048–5059.

Zhu H., Parker W., Basnar R., Proracki A., Falletta P., Béland M. and Seto P. (2008). Biohydrogen production by anaerobic co-digestion of municipal food waste and sewage sludges. *International Journal of Hydrogen Energy*, **33**(14), 3651–3659.

Zuo Z., Wub S., Zhang W. and Dong R. (2013). Effects of organic loading rate and effluent recirculation on the performance of two-stage anaerobic digestion of vegetable waste. *Bioresource Technology*, **146**, 556–561, http://dx.doi.org/10.1016/j.biortech.2013. 07.128

5

TEs in AD Systems: Major Physicochemical Processes

Bikash Chandra Maharaj[1,2,4]*, Maria Rosaria Mattei[2], Luigi Frunzo[2], Artin Hatzikioseyian[3], Eric D. van Hullebusch[4,5], Piet N. L. Lens[4] and Giovanni Esposito[6]

[1]*University of Cassino and the Southern Lazio, Department of Civil and Mechanical Engineering, Cassino, Italy*
[2]*University of Naples Federico II, Department of Mathematics and Applications 'Renato Caccioppoli', Naples, Italy*
[3]*Laboratory of Environmental Science and Engineering, School of Mining and Metallurgical Engineering, National Technical University of Athens (NTUA), 9 Heroon Polytechniou, 15780, Athens, Greece*
[4]*IHE Delft Institute for Water Education, Department of Environmental Engineering and Water Technology, The Netherlands*
[5]*Institut de Physique du Globe de Paris, Sorbonne Paris Cité, Université Paris Diderot, UMR 7154, CNRS, F-75005 Paris, France*
[6]*University of Naples Federico II, Department of Civil, Architectural and Environmental Engineering, Naples, Italy*

*Corresponding author: Department of Civil and Mechanical Engineering, University of Cassino and Southern Lazio, Via di Basio 43, 03043 Cassino, Italy

ABSTRACT

Trace elements (TEs) are essential for microbial activity in anaerobic environments. They are often added to improve the biogas production rate and yield. Dosing of TEs in anaerobic digestion (AD) systems is largely based on trial-and-error approach as no general guidelines exist to date. This is primarily because the fate of TEs in AD environments still remains poorly understood. This knowledge gap is due to the multiple and complex biogeochemical processes influencing TEs chemistry, TE physicochemical interactions with biotic and abiotic surfaces, as well as uptake by microbial community. A mathematical model based on TE-dosing optimization can be recruited to tackle such a complex problem. In this regard, the major physicochemical processes involved in determining the fate of TEs in an anaerobic-digestion environment need to be reviewed and consolidated with a suitable modelling approach. This chapter enlists and describes the most important physicochemical processes such as precipitation, adsorption, and aqueous complexation, as well as the bio-uptake mechanisms involving TEs in AD systems with the aim of summarizing the main modelling contributions to determine the fate of TEs in engineered anaerobic-digestion environments.

KEYWORDS: modeling, anaerobic digestion, trace elements

5.1 INTRODUCTION

Nutrients are essential for living organisms to carry out catabolic and anabolic biochemical reactions. The microbial requirement in terms of nutrients has been manipulated in all bioprocess technologies (England, 2013). Nutrients can be classified in two broad categories: macro- and micro- nutrients. Macronutrients are the bulk of the energy molecules required by the microbes, while micronutrients support the metabolism in the form of structural molecules. Micronutrients also enhance the rate and specificity of biochemical reactions. Accordingly, balanced macro- and micronutrients are required for ideal growth conditions and are essential for efficient and stable biogas production (Fermoso et al., 2008). Any imbalance in these factors can affect the activity and syntrophy of microorganisms in an anaerobic digester (Gustavsson, 2012; Jiang, 2006) and hence limit biogas production. The presence of oxygen, the accumulation of volatile fatty acids (VFAs) and the presence of toxic compounds/elements (i.e. ammonium, high concentrations of TEs) are the major disturbances reported for AD systems (Chen et al., 2008). Among all the microbial communities prevailing in an anaerobic environment, methanogens are the most sensitive to environmental perturbations and hence their activity is easily disturbed.

Table 5.1 lists the most important micronutrients or trace elements (TEs) involved in AD systems and the stimulating concentration range reported in some experimental studies performed with pure microbial strains (Glass & Orphan, 2012). In this regard, Fe, Co an Ni have been recognized as essential micronutrients (Glass & Orphan, 2012; Oleszkiewicz & Sharma, 1990; Thanh et al., 2015; Uemura, 2010). The effect of TEs on AD has been extensively reported (Fermoso et al., 2008; Gonzalez-Gil & Kleerebezem, 1999; Jiang, 2006; Mudhoo & Kumar, 2013; Roussel, 2012; Thanh et al., 2015; Zandvoort et al.,

2006). Cobalt forms the core of the vitamin B12 which acts as a cofactor for the enzyme methylase involved in methane production (Banerjee & Ragsdale, 2003; Stupperich & Kräutler, 1988; Mazumder *et al.*, 1987). The coenzyme F430 containing Ni is essential for the functioning of the methylcoenzyme M reductase, which is involved in the reduction of coenzyme M to methane in methanogens (Finazzo *et al.*, 2003; Thauer, 1998). Iron is involved in the transport system of the methanogenic archaea for the conversion of CO_2 to CH_4, where it functions both as an electron acceptor and donor (Thanh *et al.*, 2015), and also binds to sulfide to form precipitates which in turn reduce the hydrogen sulfide content of the anaerobic digester (Gustavsson, 2012; Hille *et al.*, 2004; Kong *et al.*, 2016; Oude Elferink *et al.*, 1994; Shakeri *et al.*, 2012). Similarly, Mn acts as an electron acceptor in anaerobic processes. Zn, Cu, and Ni have all been found in hydrogenase (Thanh *et al.*, 2015). Trace elements such as W and Mo are also found in enzymes such as formate dehydrogenase involved in formate formation from propionate by propionate oxidizers (Thanh *et al.*, 2015). Tungsten can also be considered as essential, but its low concentration in the sludge in comparison with the other metals (Fe, Ni and Co) reduces the interest for supplementing this metal (Fermoso *et al.*, 2009; Oleszkiewicz & Sharma, 1990).

Table 5.1 List of TEs with optimal dissolved concentration in growth media for pure methanogenic cultures grown on different substrates (adapted from Glass & Orphan, 2012).

Metal	Substrate	Concentration (µM)	Species
Fe	H_2/CO_2	300–500	*Methanospirillum hungatei*
	H_2/CO_2	>15	*Methanococcus voltae*
	Acetate	100	*Methanothrix soeggenii*
	Methanol	50	*Methanosarcina bakeri*
Ni	H_2/CO_2	1	*Methanobacterium thermoautotrophicum*
	H_2/CO_2	0.2	*Methanococcus voltae*
	Acetate	2	*Methanothrix soehngenii*
	Methanol	0.1	*Methanosarcina barkeri*
Co	Acetate	2	*Methanothrix soehngenii*
	Methanol	1	*Methanosarcina barkeri*
Mo	Acetate	2	*Methanothrix soehngenii*
	Methanol	0.5	*Methanosarcina barkeri*
W	H_2/CO_2	1	*Methanocorpusculum parvum*
	Formate	0.5	*Methanocorpusculum parvum*

Within biological systems, the physicochemical processes are those that are not directly mediated by microbial activity, but affect the biochemical processes to a large extent (Batstone *et al.*, 2012). Physicochemical processes including precipitation, co-precipitation, adsorption, surface complexation and gas production are among the most important taking place in AD. These processes occur in the extracellular environment and affect the overall performance of the system by mediating a change in the proton balance (hence in pH) of the bioreactor (Fermoso *et al.*, 2015). In particular, the physicochemical processes are directly related to the fate of TEs in AD and, therefore, to the bioavailability of TEs for bio-uptake. Some of the main physicochemical mechanisms related to TEs in AD are presented in Figure 5.1.

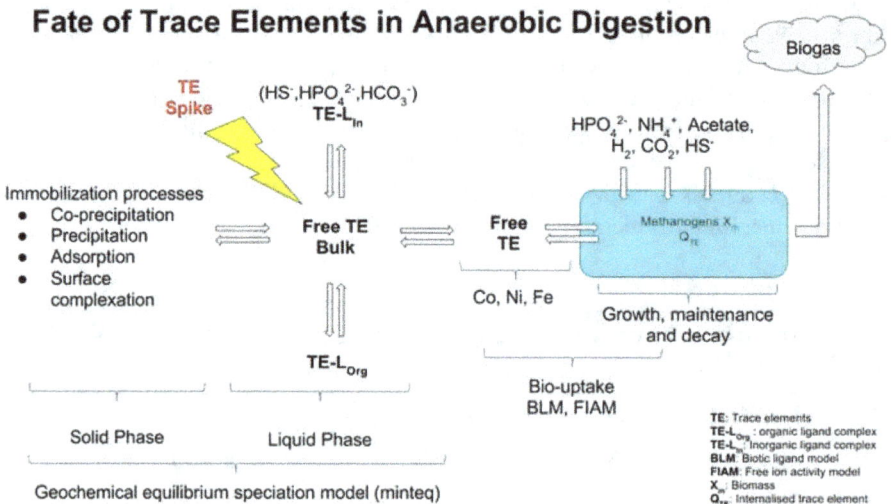

Figure 5.1 Schematic representation of various processes affecting the dynamics of TEs in an engineered anaerobic system.

The underlying principles of the physicochemical processes are relatively well understood (Batstone *et al.*, 2012; Stumm & Morgan, 1996). However, existing AD models, including the Anaerobic Digestion Model 1 (ADM1) as the most representative (Batstone *et al.*, 2002), lack a thoughtful implementation of many of those processes occurring in AD (Batstone *et al.*, 2012; Fermoso *et al.*, 2015; Flores-Alsina *et al.*, 2016; Maharaj *et al.*, 2018; Mbamba *et al.*, 2015a, b; Xu *et al.*, 2015; Zhang *et al.*, 2015).

The understanding and efforts to model and quantify the fate of TEs in AD are isolated and patchy. This chapter attempts to consolidate the available knowledge and efforts on modelling the fate of TEs in AD and envisages a set of processes necessary to quantify the TE dynamics. The chapter presents the principles and methodologies to incorporate the effect of TEs in the structure of AD models. Figure 5.2 summarizes a conceptual interaction diagram between methane-producing microbes, sulfate-reducing bacteria and TEs. It is clearly depicted that the effect of TEs should be studied in an integrated competing biotic and abiotic environment in an AD model. For all the processes to be discussed in the following sections, the relative models have been classified in two broad categories: the first corresponds to the mathematical models that have been formulated for the specific process independently from the consideration of an AD environment, while the second contemplates the models that can be considered to be an extension of the original ADM1 model. This option is justified by the significant impact that the ADM1 model has on AD literature. Indeed, in the timeline of the published models, ADM1 is considered a significant milestone (Batstone *et al.*, 2002). ADM1 was the result of a collaborative work from the International Water Association (IWA) task group for mathematical modelling of AD processes. The authors tried to unify and consolidate the knowledge and modelling experience accumulated in the literature until 2002, when the scientific and technical report of ADM1 was first published (Batstone *et al.*, 2002). ADM1 is considered even today a state-of-the-art model, including a significant number of biological, chemical and physicochemical processes (Batstone *et al.*, 2002). The model was conceived as a general modelling framework for AD processes and as such has been used successfully without modifications (Blumensaat & Keller, 2005) or in a modified version (Galí *et al.*, 2009; Parker & Wu, 2006; Peiris *et al.*, 2006) by many researchers to simulate experimental results. ADM1 has also been used as a complementary module to activated sludge models such as activated sludge model 1 (ASM1), activated sludge model 2d (ASM2d) or activated sludge model 3 (ASM3) for plant-wide wastewater treatment processes (Copp *et al.*, 2003; Kauder *et al.*, 2007; Lopez-Vazquez *et al.*, 2013; Nopens *et al.*, 2009; Rosen *et al.*, 2006).

Note that the majority of mathematical models under consideration in this work do not explicitly consider TEs as model components. However, such models are discussed as they introduce a general framework that could be extended/adapted to the case of TEs in AD systems. Furthermore, as the mathematical modelling of the fate of TEs in AD systems cannot neglect the explicit consideration of sulfur and phosphorus dynamics, which play a crucial role in precipitation and adsorption processes, the main mathematical models accounting for sulfur and phosphorus in anaerobic environments have been reviewed as well.

This chapter is organized as follows: in Section 5.2 the main models for sulfur and phosphorus dynamics in anaerobic environments are described. In Section 5.3 the main physicochemical processes affecting TE availability in AD systems are described and the modelling approaches proposed over the years are reviewed and summarized. Specifically, Section 5.3.1 is dedicated to precipitation modelling; in Section 5.3.2 adsorption mechanisms and the main modelling approaches to this topic are presented; Section 5.3.3 is devoted to aqueous complexation and the relative modelling approaches. The chapter has been extended with Section 5.4

which refers to the mathematical modelling of TEs bio-uptake and reviews the mathematical formulas usually adopted for dose-response functions.

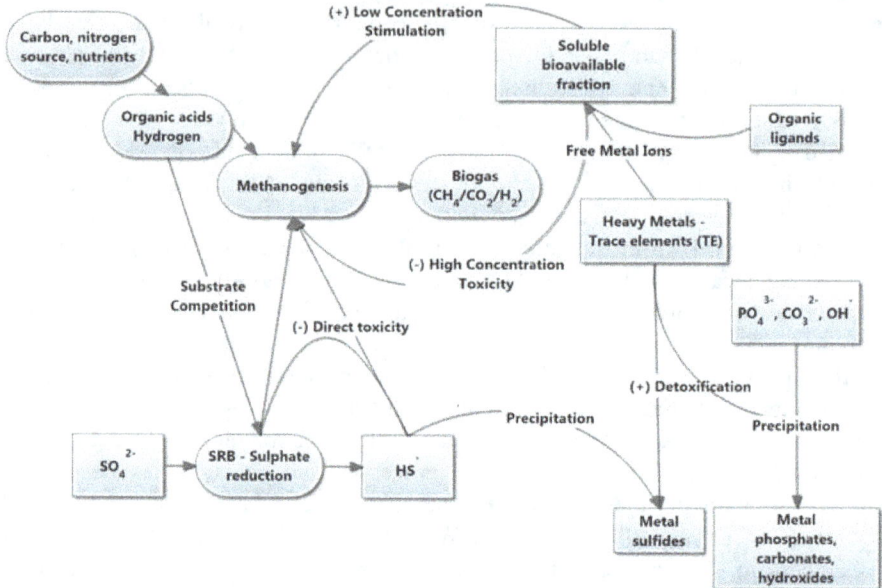

Figure 5.2 Conceptual interactions between methanogenic activity, sulfate reduction and trace elements positive (+) or negative (−) effects on biogas production.

5.2 MATHEMATICAL MODELLING OF SULFUR AND PHOSPHORUS CYCLES IN ANAEROBIC DIGESTION SYSTEMS

5.2.1 Sulfur modelling

Sulfur represents a dominant element in anaerobic digesters, counting on multiple sources. In particular, sulfide precipitates have been identified as one of the major sinks for sulfur-related compounds and subsequent removal of heavy metals toxicity (Lawrence & McCarty, 1965). Due to the significant participation of sulfur compounds in physicochemical processes affecting TEs fate and bioavailability in AD systems, it is necessary to consider sulfur transformations such as sulfate reduction in AD models (Paulo et al., 2015). In this regard, ADM1 lacks the representation of sulfur bioprocesses. The extension of any AD models with sulfate-reduction processes is considered prerequisite when TEs are also included in the system. This is related to the fact that most TEs form stable insoluble solid sulfide precipitates under sulfidogenic conditions. This process significantly affects the bioavailable fraction of TEs and cannot be ignored (Paulo *et al.*, 2015). Nonetheless, some studies have been carried out for modelling sulfate reduction in AD. Sulfate-reduction modelling necessitates the definition of important biochemical, gas transfer, and physicochemical process rates to link new state variables. These processes and state variables are mentioned along with the modelling efforts discussed in the following sections.

Sulfate-reduction models have been developed and calibrated for specific purposes for different substrates and for simulating specific systems. Some follow a different principle (other than ADM1) in their conception and hence they are classified in this chapter as 'stand-alone sulfate-reduction models' (Kalyuzhnyi & Fedorovich, 1997, 1998; Knobel & Lewis, 2002; Poinapen & Ekama, 2010; Ristow *et al.*, 2002; Vavilin *et al.*, 1995). Conversely, models which are embedded in the structure of ADM1 have been termed as 'ADM1-based sulfate-reduction models' (Barrera *et al.*, 2015; Federovich *et al.*, 2003; Flores-Alsina *et al.*, 2016). These two groups of models are discussed in the following sections.

5.2.1.1 Stand-alone sulfate-reduction models

Sulfate-reduction models have evolved independently as well as part of more complex AD models. These models do not follow the ADM1 structure as the theoretical basis, for some of them were already set before the inception of ADM1. In an attempt to model the physicochemical reaction system for pH prediction, a simulation model for AD was developed (Vavilin *et al.*, 1995). The model kept track of the sulfide formation and its effect on pH. It also considered two sulfate-reducing microbial groups which separately use acetate and propionate as organic substrate to produce sulfide from sulfate. A reduced model of self-oscillating dynamics in an AD system with sulfate reduction was developed in (Kalyuzhnyi & Fedorovich, 1998). The model considered only two groups of microbes, methanogens and sulfate reducers. Substrate-limiting functions were described by typical Monod equations. The model considered hydrogen sulfide inhibition as a function of non-ionized hydrogen sulfide. Finally, the model was validated with experimental data. The competition between sulfate-reducing bacteria (SRB) and methanogens by using data from a synthetic high sulfate-containing wastewater treated in a UASB reactor was evaluated in (Kalyuzhnyi & Fedorovich, 1998). There was a good correlation between model simulations and experimental data. Competition for acetate was emphasized. The modelling results were obtained under variations of hydraulic retention time (HRT), SO_4^{2-}/COD ratio, initial proportion of SRB/methanogens in seed sludge, efficiency of retention of SRB, and sludge retention time. Similarly, the competition between methanogens and SRB in a UASB reactor was studied to develop a dispersed plug-flow model (Kalyuzhnyi *et al.*, 1998). Data from literature were used to calibrate the model with emphasis on acetate competition. The model was able to predict system performance with regard to variations in the liquid upward velocity as an important control parameter. However, when pH changes occurred the results were not correctly predicted. The model of Knobel and Lewis (2002) was calibrated and validated for a number of reactor configurations: packed bed, UASB and gas lift reactor. Three simulation studies were carried out in steady-state and dynamic conditions. The first two simulation tests were able to predict concentration of sulfate and COD in the effluent. AQUASIM was used (Ristow *et al.*, 2002) in a recycling sludge bed reactor to simulate the AD process including sulfate reduction. Acid-mine drainage and primary sludge were used as sulfate-containing wastewater and carbon source, respectively. Data from pilot plant agreed well with model simulations. Influence of sludge recycle ratio, SO_4^{2-}, COD ratio and HRT on sulfate-reduction process was studied. A two-phase (aqueous-gas) kinetic model for sulfate-reduction processes using primary sewage sludge as carbon source was developed in Poinapen & Ekama 2010. The kinetic model was calibrated with experimental data starting from different SO_4^{2-}/COD ratios (Poinapen & Ekama 2010). Model simulations showed better fitting with experimental data. In addition, model simulations showed that at ambient temperature of 20°C, the hydrolysis rate is significantly reduced as compared with 35°C. The hydrolysis rate of the primary sewage sludge biodegradable particulate organism is the same under methanogenic and sulfidogenic conditions. The primary sewage sludge and biodegradable particulate organics are carbon deficient for biological sulfur reduction. Moreover, the model only considered gas stripping for H_2S. The efficiency of retention of SRB, HRT, SO_4^{2-}, butyrate, propionate, and acetate were predicted. However, the removal efficiencies for COD, SO_4^{2-}, hydrogen sulfide concentrations in the gas phase were not predicted (Barrera *et al.*, 2013). In summary, 'stand-alone sulfur-reduction models' have been developed and calibrated to study the impact of operating parameters and initial concentrations under specific cases.

5.2.1.2 ADM1-based sulfate-reduction models

Sulfate-reduction processes as an extension of ADM1 have been implemented in Federovich *et al.* (2003). The model was calibrated against experimental data from literature and was able to predict sulfate removal in the AD process, the concentrations of butyrate, propionate and acetate, as well as methane and biomass production. The extended model included sulfate reduction and biogenic sulfide production by SRB. SRB compete with methanogens for intermediates produced during the biodegradation processes. The extension added four different SRB species ($X_{SRB\ butyrate}$, $X_{SRB\ propionate}$, $X_{SRB\ acetate}$, $X_{SRB\ hydrogen}$) which oxidize butyrate, propionate, acetate and hydrogen, respectively, coupled with sulfate reduction according to the following reactions:

$$2C_3H_7COO^- + SO_4^{2-} \xrightarrow{X_{SRB\ butyrate}} 4CH_3COO^- + H^+ + HS^- \tag{5.1}$$

$$4C_2H_5COO^- + 3SO_4^{2-} \xrightarrow{X_{SRB\ propionate}} 4CH_3COO^- + 4HCO_3^- + H^+ + 3HS^-$$
$$(5.2)$$

$$CH_3COO^- + SO_4^{2-} \xrightarrow{X_{SRB\ acetate}} 2HCO_3^- + HS^- \qquad (5.3)$$

$$4H_2 + H^+ + SO_4^{2-} \xrightarrow{X_{SRB\ hydrogen}} HS^- + 4H_2O \qquad (5.4)$$

The following rate equations for reactions (5.1)–(5.4) have been adopted:

$$r_i = k_{max,i} \frac{S_i}{K_{S_i} + S_i} \cdot \frac{S_{SO_4}}{K_{SO_4} + S_{SO_4}} \cdot X_i \cdot I_{pH} \cdot I_{sulfide} \qquad (5.5)$$

where r_i is the volumetric growth rate of SRB group X_i [M L^{-3} T^{-1}], $k_{max,i}$ is the maximum specific growth rate for group X_i [T^{-1}], S_i represents the concentration of substrate i [M L^{-3}], K_{S_i} denotes the Monod half-saturation constant for substrate i [M L^{-3}], S_{SO_4} is the concentration of sulfate in liquid phase [M L^{-3}], K_{SO4} represents the Monod half-saturation constant for sulfate [M L^{-3}], X_i is the concentration of SRB biomass group i [M L^{-3}], I_{pH} is the pH inhibition factor (dimensionless), and $I_{sulfide}$ represents the sulfide inhibition factor (dimensionless). The inhibition by hydrogen sulfide was included in the rate equation (5.5) through the term:

$$I_{sulfide} = 1 - \frac{S_{H_2S}}{K_I} (if\ S_{H_2S} > K_I, I_{sulfide} = 0) \qquad (5.6)$$

where S_{H_2S} denotes the concentration of undissociated hydrogen sulfide in liquid phase [M L^{-3}]; and K_I is the inhibition constant of undissociated hydrogen sulfide [M L^{-3}].

The values of the kinetic parameters for SRB were estimated in the model to simulate the behavior of a UASB fed with sulfate up to 6 g S/L (Federovich et al., 2003). Although this model is reported today as the most appropriate extension of ADM1 when sulfate-removal efficiencies are of primary interest, it was not calibrated to predict the concentrations of total aqueous sulfide, undissociated sulfides and gas phase sulfides.

Similarly, an extension of ADM1 with sulfate reduction was proposed in (Barrera et al., 2015). The model was calibrated and validated for a high strength and sulfate rich wastewater (cane molasses vinasse). Butyric acid was neglected as substrate for SRB in the model structure. Propionic acid, acetic acid, hydrogen were considered as primary electron donors for sulfate reduction processes. Propionate SRB, acetate SRB and hydrogenotrophic SRB were considered in addition to the seven microbial groups of ADM1. The new biochemical reactions introduced in the model have been depicted in the following equations:

$$C_2H_5COOH + 0.75H_2SO_4 \xrightarrow{X_{SRB\ propionate}} CH_3COOH + CO_2 + H_2O + 0.75H_2S$$
$$(5.7)$$

$$CH_3COOH + H_2SO_4 \xrightarrow{X_{SRB\ acetate}} 2CO_2 + 2H_2O + H_2S \tag{5.8}$$

$$4H_2 + H_2SO_4 \xrightarrow{X_{SRB\ hydrogen}} H_2S + 4H_2O \tag{5.9}$$

A dual Monod type kinetic was used for the uptake of the substrates. Protonation and deprotonation of sulfuric acid and sulfide ions were considered in the liquid phase. Gas stripping of H_2S was also considered. S_{H_2S} was included as the process inhibitor. Likewise, a non-competitive inhibition function for sulfides was introduced. The model prediction works reasonably well for the process variables with an accurate quantitative predictions of high ($\pm 10\%$) to medium (10%–30%) and error ranging from 1 to 26%.

Recently in (Flores-Alsina *et al.*, 2016), biological production of sulfide by sulfate reducing bacteria has been included as one of the four extensions to ADM1, the others being phosphorus metabolism, mineral precipitation and iron transformation in an AD system. The model was applied to a set of full-scale plant-wide data.

Selected process rates and inhibition functions, which may be used to model sulfate reduction processes, are represented in the equations below. Readers may refer to the literature for more details (Barrera *et al.*, 2013).

Growth rate for SRB species:

$$r_i = k_{max,i} \frac{S_i}{K_{S_i} + S_i} \cdot \frac{S_{SO_4}}{K_{SO_4} + S_{SO_4}} \cdot X_i \cdot I_{pH} \cdot I_{H_2S} \tag{5.10}$$

where, r_i is the volumetric growth rate of SRB group $i[ML^{-3}T^{-1}]$; $k_{max,i}$ is the maximum specific growth rate of SRB group $i[T^{-1}]$; S_i is the concentration of soluble organic components $[ML^{-3}]$; K_{S_i} is half saturation constant $[ML^{-3}]$; X_i is the concentration of SRB group $i[ML^{-3}]$; I_{pH} is the pH inhibition function; and I_{H_2S} is the sulfide inhibition function.

Biomass decay rate:

$$r_{decay,i} = k_{dec,i} \cdot X_i \tag{5.11}$$

where, $\rho_{decay,i}$ is the kinetic rate of substrate uptake (kg COD m$^3 d_{-1}$); and $k_{dec,i}$ is first order decay rate of species $i(d^{-1})$;

Acid-base rate for sulfide species:

$$\rho_{A/B} = K_{A/B,H_2S}(S_{HS^-} \cdot (S_{H^+} + K_{a,H_2S}) - K_{a,H_2S} \cdot S_{H_2S,tot}) \tag{5.12}$$

where $\rho_{A/B}$ is the acid-base kinetic rate $[ML^{-3}T^{-1}]$; $k_{A/B,H_2S}$ is the acid-base kinetic parameter $[M^{-1}L^3T^{-1}]$; S_{HS^-} is the concentration of bisulfide ion $[ML^{-3}]$; S_{H^+} is the concentration of hydrogen ion $[ML^{-3}]$; $S_{H_2S,tot}$ is the concentration of total H_2S species (dissociated and undissociated) in liquid phase $[ML^{-3}]$; and K_{a,H_2S} is the acid-base equilibrium coefficient $[ML^{-3}]$.

Gas transfer:

$$\rho_{gas} = k_l a \cdot (S_{H_2S} - K_{H,H_2S} \cdot P_{gas,H_2S}) \tag{5.13}$$

where ρ_{gas} is the specific mass transfer rate of gas $H_2S[ML^{-3}T^{-1}]$; $k_l a$ is the gas-liquid mass transfer coefficient $[T^{-1}]$; S_{H_2S} is the concentration of undissociated $H_2S[ML^{-3}]K_{H,H_2S}$ is Henry's law coefficient $[M^2L^{-3}T^{-1}]$; and P_{gas,H_2S} is the partial pressure of the H_2S gas $[MT^{-1}]$.

Process inhibition:

Sulfide:

$$I_{H_2S} = e^{-\left(\frac{S_{H_2S}}{K_{I,H_2S}}\right)^2} \tag{5.14}$$

where K_{I,H_2S} is the inhibition coefficient by undissociated $H_2S[ML^{-3}]$.

$$I_{H_2S} = \frac{1}{1 + \dfrac{S_{H_2S}}{K_{I,H_2S}}} \tag{5.15}$$

$$I_{H_2S} = \frac{1}{1 + \left(\dfrac{S_{H_2S}}{K_2}\right)^{\ln99/\left(\frac{K_{100}}{K_2}\right)}} \tag{5.16}$$

where K_{100} is the concentration of H_2S or pH at which the uptake rate is decreased 100 times $[ML^{-3}]$; and K_2 is the concentration of H_2S or pH at which the uptake rate is decreased 2 times $[ML^{-3}]$.

pH:

$$I_{pH} = \frac{1}{1 + \dfrac{pH^{\ln99/(K_{100}/K_2)}}{K_2}} \tag{5.17}$$

where pH is the resultant or predicted pH of the system.

$$I_{pH} = (1 + \exp(-\alpha_{LL}(pH - pH_{LL})))^{-1} \cdot (1 + \exp(-\alpha_{UL}(pH - pH_{UL})))^{-1} \tag{5.18}$$

where α_{LL} and α_{UL} are the parameters which affect the steepness of the curve; pH_{LL} is the lower pH limit where microbial growth is 50% inhibited; and pH_{UL} is the upper pH limit where microbial growth is 50% inhibited.

$$I_{pH} = \frac{1 + 2 \cdot 10^{0.5(pH_{UL}-pH_{LL})}}{1 + 10^{(pH-pH_{UL})} + 10^{(pH_{LL}-pH)}} \tag{5.19}$$

5.2.2 Phosphorus modelling

Phosphorus is one of the most abundant inorganic components of AD systems (~30% nutrient load according to Johnson and Shang (2006)). Phosphorus transformations are major drivers for precipitation processes (struvite precipitation). Although phosphorus precipitation has been addressed in equilibrium- and kinetic-based models, phosphorus transformations have not been linked to the biodegradation reactions in the majority of studies. Poly-hydroxy-alkanoates (X_{PHA}), poly-phosphate (X_{PP}), and phosphate ($S_{PO_4^{3-}}$) are the major biochemical components which are necessary to define phosphorus transformations. Apart from this, inorganic phosphate-water acid/base system ($S_{PO_4^{3-}}/S_{HPO_4^{2-}}/S_{H_2PO_4^-}/S_{H_3PO_4}$) should be taken into account to track the change in pH. Additionally, based on requirements, inorganic phosphate minerals can also be introduced, such as, amorphous calcium phosphate ($X_{Ca_3(PO_4)_2}$), hydroxylapatite ($X_{Ca_5(PO_4)_3(OH)}$), octacalium phosphate ($X_{Ca_8H_2(PO_4)_6}$) and, struvite ($X_{MgNH_4PO_4}$). The following lines briefly discuss efforts to model phosphorus transformations in waste-treatment models and ADM1-based models.

Modelling the effect of pH change (due to phosphate species) and tracking the dynamics of phosphate minerals in relation to the biodegradation are two important aspects of modelling phosphorus in AD systems. The pH prediction module has been expanded to take into account physicochemical effects such as ion-activity correction, ion-pairing behavior, and weak acid-base reactions occurring in conjugation with biological reactions (Solon et al., 2015). Phosphorus metabolism has been implemented in the framework of ADM1 in (Johnson & Shang, 2006). Two major methods for incorporating phosphorus in the AD models have been identified (Solon, 2015). In the first approach, P does not participate in biological reactions. The quantity of phosphorus within the system, i.e. in the particulate and soluble components, is tracked through the introduction of a new state variable S_{IP} (concentration of inorganic phosphorus). This acts as the only source and sink for all the inorganic phosphorous minerals and acid-base species (Zaher & Chen, 2006; Zaher et al., 2007). In addition, it is assumed that there are instantaneous phosphorus-related processes taking place at the interface between the activated sludge and the AD process. The prevalence of inorganic phosphorus and absence of biological phosphorus oversimplifies the model structure.

Conversely, the alternate method takes into account some of the processes which are considered in the IWA Activated Sludge Model No. 2d (ASM2d) (Henze et al., 1999). These processes are further incorporated in ADM1. Phosphorus metabolizing microorganisms are active in the anaerobic digester and facilitate microbial conversion of phosphate-related compounds. Based on this principle, a model was developed in (Wang et al., 2016). The model investigated anaerobic fermentation of enhanced biological phosphorus removal of sludge in terms of phosphate release and VFA production with respect to polyphosphate-accumulating organism (PAO) activity. The proposed model extension included: (a) the storage of 4 VFA as PHA mediated by PAOs; (b) the effect of PHA content on disintegration rate; and (c) the mineral phosphate precipitation. Similar to this study, a new extension was implemented which enabled the release of phosphorus in AD (Flores-Alsina et al., 2016). For this, phosphorus-accumulating organism (X_{PAO}), polyhydroxyalkanoates (X_{PHA}) and polyphosphates (X_{PP}) constitute the new state variables. Seven new processes were added for: (a) uptake of valerate, butyrate, propionate and acetate to form polyhydroxyalkanoates (X_{PHA}); (b) decay of phosphorus accumulating organism (X_{PAO}); and (c) lysis of polyhydroxyalkanoates (X_{PHA}) and polyphosphates (X_{PP}). Further details of the model can be found in the literature (Flores-Alsina et al., 2016).

5.3 MATHEMATICAL MODELLING OF PHYSICOCHEMICAL PROCESSES IN ANAEROBIC DIGESTION SYSTEMS

Despite several efforts devoted to the improvement of physicochemical modelling and simulation in anaerobic environments, the effect and fate of TEs have not been accounted for (Fermoso et al., 2015; Lauwers et al., 2013; van Hullebusch et al., 2016). As mentioned, the most important processes to be included in the AD model in order to estimate the fate of TEs are: precipitation, aqueous complexation and surface complexation/adsorption (Batstone et al., 2012; van Hullebusch et al., 2016). A modelling approach is envisioned that takes into account all these intermittent processes and unifies them in the form of a defined model structure (Figure 5.2 above). In the last decades, there have been efforts in wastewater-treatment process modelling to predict the physico-chemical state of the system. A list of such efforts has been provided in Table 5.2. Some of these works are discussed in the following lines. In addition, the ADM1 extensions considering the mentioned processes are also described.

5.3.1 Precipitation

With the term 'non ADM1-based precipitation models', we refer to the class of models that have been developed to simulate the precipitation process in aqueous environments, non-necessarily anaerobic. In some cases such models account for the biodegradation reactions occurring in AD but they do not follow the modelling framework introduced in ADM1. Conversely, we define 'ADM1-based precipitation models', the ones that have been conceived in the framework of ADM1 and explicitly take into account precipitation reactions not necessarily for TEs. Table 5.3 represents a potential list of precipitation reactions which can occur in an anaerobic environment.

Table 5.2 Waste-treatment models considering various processes involving geochemical components.

No.	Reference	Process	Model Type	Geochemical Components	Application	
1	Hanhoun et al. (2011)	Precipitation	Equilibrium	Mg^{2+}, NH_4^+, PO_4^{3-}, H_2O	Phosphorus recovery	Lab scale
2	Scott et al. (1991)	Precipitation	Equilibrium	Mg^{2+}, NH_4^+, PO_4^{3-}, H_2O	Waste water stream, kidney stones	Lab scale
3	Barat et al. (2011)	Precipitation	Equilibrium, Kinetic	Mg^{2+}, NH_4^+, PO_4^{3-}, H_2O, CO_3^{2-}, Ca^{2+}, K^+	SBR for biological phosphorus removal	Lab scale
4	Musvoto et al. (2000b)	Acid-base reactions, precipitation, gas stripping, ion pairing	Kinetic	H^+, OH^-, H_2CO_3, HCO_3^-, CO_3^{2-}, H_3PO_4, $H_2PO_4^-$, HPO_4^{2-}, PO_4^{3-}, NH_4^+, Ca^{2+}, Mg^{2+}	Aeration of anaerobic digestion liquor from spent wine UASB and AD for sewage sludge	Plant wide
5	Maurer & Boller (1999)	Precipitation	Kinetic	H_2CO_3, HCO_3^-, CO_3^{2-}, OH^-, H^+, H_3PO_4, $H_2PO_4^-$, HPO_4^{2-}, PO_4^{3-}, Ca^{2+}	Waste treatment-EBPR	Lab scale
6	Ohlinger et al. (1998)	Precipitation	Equilibrium	Mg^{2+}, PO_4^{3-}, HPO_4^{2-}, OH^-, H^+, NH_4^+	Waste treatment	Plant wide
7	Smith et al. (2008)	Precipitation, adsorption, co-precipitation	Equilibrium	H^+, PO_4^{3-}, Fe^{2+}	Phosphate removal	Lab scale
8	Wrigley et al. (1992)	Precipitation	Equilibrium	Mg^{2+}, NH_4^+, PO_4^{3-}, H_2O	Waste water stream, kidney stones	Lab scale
9		Precipitation	Kinetic		Waste treatment	Lab scale

#	Reference	Processes	Modelling	Components	Application	Scale
	Van Rensburg et al. (2003)					
10	Tait et al. (2009)	Precipitation	Equilibrium	H^+, OH^-, H_2CO_3, HCO_3^-, CO_3^{2-}, H_3PO_4, $H_2PO_4^-$, HPO_4^{2-}, PO_4^{3-}, NH_4^+, Ca^{2+}, Mg^{2+}	Low cost sulfate removal	Lab scale
11	Hauduc et al. (2015)	Acid-base; chemical ion pairing; precipitation; chemical surface complexation; adsorption; co-precipitation	Kinetic	Multiple components	Chemical phosphorus removal	Lab scale
12	Schwarz & Rittmann (2007)	Acid-base reactions, Precipitation, surface complexation, biodegradation	Equilibrium, Kinetic,	HS^-, S^-, H^+, Zn^{2+}	Metal detoxification in sulfidic systems	Lab Scale
13	Mbamba et al. (2015b)	Acid base, precipitation-dissolution, gas stripping	Equilibrium, Kinetic	H^+, NH_4^+, Ca^{2+}, Mg^{2+}, PO_4^{3-}, PO_4^{3-}	Calcite precipitation in synthetic aqueous solution	Lab scale
14	Flynn et al. (2014)	Surface complexation	Equilibrium	Co, Ni, Sr, Zn	Calculation of metal-carboxyl bacterial surface stability constant	Lab scale
15	VanBriesen & Rittmann (1999)	Acid-base, complexation and biodegradation	Equilibrium, Kinetic	H^+, NH_4^+, OH^-, Fe^{3+}, H_2CO_3	Citrate/Fe(III) systems	Theoretical

Table 5.3 Probable precipitation reactions in AD systems.

No.	Name	Precipitation Reaction	$\text{Log}K_{sp}$	Citation in AD systems
1	Al_2O_3	$Al_2O_3 + 6H^+ \rightleftharpoons 2Al^{3+} + 3H_2O$	19.6	Mu et al. (2011)
2	$AlPO_4$	$AlPO_4 \rightleftharpoons Al^{3+} + PO_4^{3+}$	20.2	Peng & Colosi (2016)
3	Anhydrite	$CaSO_4 \rightleftharpoons Ca^{2+} + SO_4^{2-}$	4.6	MacConnell & Collins (2009)
4	Aragonite	$CaCO_3 \rightleftharpoons Ca^{2+} + CO_3^{2-}$	8.3	Musvoto et al. (2000b)
5	Brucite	$Mg(OH)_2 + 2H^+ \rightleftharpoons Mg^{2+} + 2H_2O$	17.1	Uludag-Demirer & Othman (2009)
6	$CaHPO_4$	$CaHPO_4 \rightleftharpoons Ca^{2+} + H^+ + PO_4^{3-}$	19.2	Van Langerak et al. (1999)
7	$Ca_4H(PO_4)_3 \cdot 3H_2O$	$Ca_4H(PO_4)_3 \cdot 3H_2O \rightleftharpoons 4Ca^{2+} + H^+ + 3PO_4^{3-} + 3H_2O$	47.9	Zhang (2010)
8	Calcite	$CaCO_3 \rightleftharpoons Ca^{2+} + CO_3^{2-}$	8.4	Mbamba et al. (2015a,b)
9	$Ca_3(PO_4)_2$	$Ca_3(PO_4)_2 \rightleftharpoons 3Ca^{2+} + 2PO_4^{3-}$	25.5	Ye et al. (2010)
10	Dolomite	$CaMg(CO_3)_2 \rightleftharpoons Ca^{2+} + Mg^{2+} + 2CO_3^{2-}$	16.5	Musvoto et al. (2000b)
11	$Fe(OH)_2$	$Fe(OH)_2 + 2H^+ \rightleftharpoons Fe^{2+} + 2H_2O$	12.8	Dezham et al. (1988)
12	FeS	$FeS + H^+ \rightleftharpoons Fe^{2+} + HS^-$	2.9	Fermoso et al. (2015)
13	Gibbsite	$Al(OH)_3 + 3H^+ \rightleftharpoons Al^{3+} + 3H_2O$	7.7	Qiao & Ho (1996)
14	Huntite	$CaMg_3(CO_3)_4 \rightleftharpoons 3Mg^{2+} + Ca^{2+} + 4CO_3^{2-}$	29.9	Musvoto et al. (2000a)
15	Hydroxyapatite	$Ca_{10}(PO_4)_6(OH)_5 + 5H^+ \rightleftharpoons 10Ca^{2+} + 6PO_4^{3-} + 5H_2O$	44.3	Maurer et al. (1999)
16	Mackinawite	$FeS + H^+ \rightleftharpoons Fe^{2+} + HS^-$	3.6	Morse & Arakaki (1993)
17	Magnesite	$MgCO_3 \rightleftharpoons Mg^{2+} + CO_3^{2-}$	7.4	Van Rensburg et al. (2003)
18	$Mg(OH)_2$	$Mg(OH)_2 + 2H^+ \rightleftharpoons Mg^{2+} + 2H_2O$	18.7	Huang et al. (2016)
19	$Mg_3(PO_4)_2$	$Mg_3(PO_4)_2 \rightleftharpoons 3Mg^{2+} + 2PO_4^{3-}$	23.2	Musvoto et al. (2000a)
20	Newberyite	$MgHPO_4 \cdot 3H_2O \rightleftharpoons Mg^{2+} + H^+ + PO_4^{3-} + 3H_2O$	18.1	Musvoto et al. (2000a)
21	Portlandite	$Ca(OH)_2 + 2H^+ \rightleftharpoons Ca^{2+} 2H_2O$	22.7	Gu et al. (2015)
22	Siderite	$FeCO_3 \rightleftharpoons Fe^{2+} + CO_3^{2-}$	10.5	Preis & Gamsjäger (2001)
23	Struvite	$MgNH_4PO_4 \cdot 6H_2O \rightleftharpoons Mg^{2+} + NH_4^+ + PO_4^{3-}$	13.2	Musvoto et al. (2000a)
24	Vaterite	$CaCO_3 \rightleftharpoons Ca^{2+} + CO_3^{2-}$	7.9	Musvoto et al. (2000a)
25	Vivianite	$Fe_3(PO_4)_2 \cdot 8H_2O \rightleftharpoons 3Fe^{2+} + 2PO_4^{3-} + 8H_2O$	37.7	Roussel & Carliell-Marquet (2016)

5.3.1.1 Non-ADM1-based precipitation models

Precipitation reactions in wastewater systems have been extensively studied in the last few decades (Barat et al., 2011; Donoso-Bravo et al., 2013; Hanhoun et al., 2011; Mbamba et al., 2015a, b; Maurer et al., 1999; Musvoto et al., 2000a, b; Rittmann et al., 2002; Tait et al., 2009; Van Rensburg et al., 2003). Generally, the precipitation studies focus on nutrient removal or recovery from wastewater streams. Recovery of phosphate is central to these studies. This is largely due to the associated risk of water eutrophication (Hauduc et al., 2015; Jimenez et al., 2015; Rahaman et al., 2014; Smith et al., 2008; Szabó et al., 2008). Some of these studies have been supplemented with process models. Different modelling techniques have been used to estimate the precipitation process and the effect of precipitation on the biochemical system (Mbamba et al., 2015b). In general, these modelling strategies are restricted either to thermodynamic equilibrium calculations or to a kinetic approach, and only a minor number of these studies adopted a combination of equilibrium and kinetic approaches to estimate the exent of precipitation (Barat et al., 2011). These approaches and the relative contributions have been listed in Table 5.4 and are discussed below.

The thermodynamic equilibrium method for estimating precipitation makes use of a well established multicomponent thermodynamic equilibrium approach (Allison et al., 1991; Morel & Morgan, 1972; Rittmann et al., 2002). Thermodynamic models consider a system comprising of components and species. Components can interact to form species, but individual species do not react to form new species. This modelling approach assumes the acid-base system to consist of a predefined number of organic/inorganic components and species. All acid-base reactions in the system are considered to be in equilibrium. The system of equation is solved by the Newton-Raphson algorithm using the equilibrium constants and the law of mass action (Mbamba et al., 2015b; Solon et al., 2015). A thermodynamic model for phosphate precipitation was developed to study the effect of pH, ionic strength and temperature on struvite solubility (Hanhoun et al., 2011). In the first step, experiments were conducted to determine the solubility product of struvite from synthetic solution at various temperatures. In the second step, an algorithm was developed to calculate the equilibrium constants. The algorithm is based on a hybrid resolution procedure which integrates a multi-genetic algorithm and the Newton-Raphson method in order to increase computing efficiency. Using this approach, the precipitation of struvite was predicted both quantitatively and qualitatively.

A predictive approach, to determine the amount of precipitate formed when arbitrary amounts of N, Mg and P were added in wastewater system, was developed by (Scott et al., 1991; Wrigley et al., 1992). The computer program used NH_3 as the central species for calculating the precipitated composition. It also included a routine to calculate the activity coefficients of each species involved. After selecting H^+ and Mg^{2+} as main species, concentration ratios (K)

Table 5.4 Representative kinetic equations for mineral-precipitation modelling.

No.	Reference	Model Equations	Notation
1	Ali & Schneider (2008)	$SI = \log\left(\dfrac{P_{SO}}{P_{cs}}\right)^{1/v}$ $G = KS^n$ $\dfrac{dM}{dt} = \dfrac{1}{2}KN\pi\rho_c L^2 S^n L^n$	where • SI is the saturation index; • P_{SO} is the concentration of total species; • P_{cs} is the conditional solubility product; • v is the number of reactants; • G is the growth rate of crystals; • K is a kinetic parameter; • n is a kinetic parameter; • N is the crystal number; • ρ_c is the density of the crystals; • L is the final size of the crystal; • S is the function of supersaturation of crystal; • M is concentration of precipitated crystal.
2	Ebigbo et al. (2012)	$SI = \dfrac{\gamma_i m_i + \gamma_j m_j}{k_{sp}}$	where • SI is the saturation index; • γ_i and γ_j are the activity coefficient of the components; • k_{sp} is the solubility product; • m_i and m_j are the concentration of the components.
3	Jeen et al. (2012)	$R_i^m = K_{eff,i}\left(\dfrac{IAP_i^m}{k_i^m}\right)^{-1}$	where • R_i^m is the rate of mineral precipitation; • $K_{eff,i}$ is the effective rate constant of precipitation; • IAP_i^m is the activity product; • k_i^m is the equilibrium constant for dissolution.

4	Mbamba et al. (2015a,b)	$$r_{cryst} = k_{cryst} X_{cryst} \sigma^n$$ $$\sigma = k_{cryst} X_{cryst} \left(\left(\frac{Z_i Z_j}{k_{sp}} \right)^{1/v} - 1 \right)^n$$	where • r_{cryst} is the rate of crystallization; • k_{cryst} is the kinetic rate constant for crystallization; • X_{cryst} is the concentration of crystal formed; • σ is the saturation index; • Z_i and Z_j are the concentrations of participating components; • v is the summation of charges; • n is an experimentally determined constant; • k_{sp} is the solubility product.
5	Rittmann et al. (2002)	$$R_p = k'_a \left(1 - \frac{k_{sp}}{Q} \right) Me$$	where • R_p is the rate of precipitation; • Me is the concentration of cationic species (i.e. TEs); • k'_a is the lumped product of reaction or mass transfer rate times; • a is the specific surface area; • k_{sp} is the solubility product; • Q is the reaction product of precipitate.
6	Koutsoukos et al. (1980)	$$\frac{d[M_{v+} A_{v-}]}{dt} = k_r \left[\left[[M^{m+}]^{v+} - [A^{a-}]^{v-} \right]^{1/v} - (k'_{sp})^{1/v} \right]^n$$	where • k_r is a kinetic constant; • $[M^{m+}]$ and $[A^{a-}]$ are the ion concentration; • $[M_{v+} A_{v-}]$ is the mineral precipitate concentration; • k'_{sp} is the solubility product; • v is the summation of opposite charges $v^+ + v^-$; • n is a constant with value 2 for sparingly soluble salts.

(Continued)

Table 5.4 Representative kinetic equations for mineral-precipitation modelling (*Continued*).

No.	Reference	Model Equations	Notation
7	Maurer et al. (1999)	$$\frac{dX_P}{dt} = k_P f_i m_i f_j m_j K_P \left(K_{PPT} + \frac{X_P}{X_{TS}} \right)^{-1}$$	where • X_P is the mineral precipitate; • m_i and m_j are the concentrations of soluble ionic components; • k_P is a kinetic constant for precipitation; • K_P is the hyperbolic coefficient; • X_{TS} is the conentration of total solids; • f_i and f_j are the activity coefficients.
9	Shaojun and Mucci (1993)	$$R = R_f - R_b = k_f (\gamma_i m_i)^{n_1} (\gamma_j m_j)^{n_2} - k_b$$	where • R_f and R_b are the forward and backward reaction rates; • k_f and k_b are the forward and backward rate constants; • m_i and m_j are the concentrations of ionic components; • γ_i and γ_j are the activity coefficients of the components; • n is the reaction rate order.

were assigned to all the equilibria followed by an approximation of the coefficients by a set of quadratic equations. Once the quadratic equations were set and solved, the next step involved the estimation of PO_4^{3-} using the charge balance (electroneutrality) of the system. Consequently, the system of equations was solved for other species, checked for consistency and the range and output plotted. The authors argued the importance of a theoretical estimation to guide further laboratory and field experimentation.

Equilibrium calculation programs have also been used to study precipitation. These programs calculate the equilibrium composition based on equilibrium constant approach. A new method to predict struvite potential in AD systems was developed by Ohlinger *et al.* (1998). This method included the ionic strength, ionic activities, magnesium phosphate complexation effects on ion speciation and the experimentally determined struvite solubility constant. The method was verified using *MINEQL+*. *MINEQL+* was used to calculate a set of theoretical equilibrium concentration values for ions and complexes present in the system. Theoretical total magnesium, total ammonia, and total phosphorus concentrations were calculated. The measured theoretical concentrations and variances from lab tests were used to produce an objective function. The objective function was minimized to arrive at a solubility constant, $pK_{SP} = 13.26$ for struvite.

The thermodynamic approach to quantify precipitation in aqueous environments generally involves the prediction of the equilibrium composition of the system at a given point in time. Such a method does not provide information about the dynamics of the species or components taking part in the precipitation reaction. Additionally, studies based on thermodynamic approach did not take into account the biological processes taking place in the system. Hence, the influence of the biological processes, in terms of pH, on precipitation and vice-versa was largely neglected.

Some studies (Barat *et al.*, 2011; Mbamba *et al.*, 2015a,b) adopted a method in which the thermodynamic and kinetic approaches were combined to model the system. In practice, some of the reactions were described by adopting the equilibrium approach whereas others processes were kinetically solved. Such a choice largely depends on the process being modelled and the precipitation reaction under study. More often, protonation/deprotonation reactions are modelled as equilibrium based while slow precipitation reactions are kinetically controlled. For instance in Barat *et al.* (2011), a calcium phosphate precipitation model was incorporated into the Activated Sludge Model No.2 (ASM2d). The model considered two types of reactions resulting in precipitation of calcium phosphate. The aqueous phase reactions (fast) and ion-paring reactions were modeled as equilibrium reactions by adopting the equilibrium approach, while the aqueous to solid-phase reactions (slow) were kinetically modelled. A surface-based kinetic was considered for precipitation reactions (solid phase). The use of a complete kinetic approach would have required one differential equation per species (28 in total), while the use of the equilibrium approach reduced the problem to 7 differential equations, one for each of the component, thus increasing the computation efficiency.

Over the course of years, full kinetic models evolved as well. The kinetic method of estimating the precipitation of mineral phases in wastewater systems assumes all the physicochemical reactions in the system as kinetic reactions in a dynamic state equation set. This set of differential equations is solved with respect to the variables describing the system. With such a method, the time varying nature of the species and components can be effectively traced and used to set up a decision-making strategy. For example, a dynamic model was formulated by Maurer & Boller (1999), based on the experimental observations that, in inactivated sludge containing high amounts of dissolved calcium and phosphorus, a pH-sensitive and reversible precipitation process exists. The model described the calcium and phosphate precipitation by using protonation/deprotonation reactions of phosphoric acid, carbonic acid and water. It also included fully reversible precipitation of hydroxydicalcium phosphate $(Ca_2HPO_4(OH)_2$, HDP) and formation of hydroxyapatite $(Ca_5(PO_4)_3OH$, HAP) from HDP. HAP is considered to be a stable product. All the reactions in the system were kinetically modeled. This approach was preferred because the effluent concentrations of the waste treatment plants are generally modeled as dynamic state variables.

The first and foremost attempt to model the three phases of an AD system, in the context of mineral precipitation, came from Musvoto et al. (2000a, b). In this study, a kinetic model for the carbonate system was developed. Ion-pairing and precipitation reactions were included in the model to describe chemical changes due to the formation of struvite, newberyite, amorphous calcium and magnesium carbonate. Further, stripping of CO_2 and NH_3 was considered. Aqueous phase mixed weak acid/base chemistry and ion-pairs constituted the ion-pair module. This was achieved by implementing a similar kinetic model for both the association/dissociation acid/base reactions and the equilibria of ion-pair formation. The aqueous phase weak acid/base submodule involved water, carbonate, phosphate, short chain fatty acid and ammonium. In total, 15 components and 26 processes were defined to implement the aqueous phase weak acid/base submodule. Likewise, to implement the ion-pair submodule, a total of 12 compounds and 22 processes were used. Here it should be noted that the forward reaction rate for ion pair formation was selected to be very high (10^7 s^{-1}) and the reverse rate was calculated subsequently from the stability constant values. As mentioned earlier, the model also included a sub-module for multiple mineral precipitation. Attention was given to magnesium, calcium and few other relevant minerals. Four possible magnesium phosphates $(MgNH_4PO_4 \cdot 6H_2O, MgHPO_4 \cdot 3H_2O, Mg_3(PO_4)_2 \cdot 8H_2O, Mg_3(PO_4)_2 \cdot 22H_2O)$, five calcium phosphates $(Ca_5(PO_4)_3OH, Ca_3(PO_4)_2, Ca_8(HPO_4)_2(PO_4)_4 \cdot 5H_2O, CaHPO_4, CaHPO_4 \cdot 2H_2O)$ and other minerals $(MgCO_3 \cdot 3H_2O)$ were considered. A concentration-based kinetic rate formulation was used with a default order of reaction of 2. Solubility product values were modified according to the Debye-Hückel rule for low and medium saline waters. The solubility product values did not consider the temperature variation. Calibration and validation of the model was carried out by experimental batch results from a spent wine upflow anaerobic sludge bed (USAB) digester and a sewage sludge digester. Further, experimental values from literature were also used to fit the model simulations. The model was implemented on the *AQUASIM* platform (Reichert 1994). The quantitative estimation of precipitation of carbonates and phosphates as well as the amount of CO_2 and NH_3 stripping from the digester was justified.

In a recent study (Mbamba *et al.*, 2015a), a set of detailed experiments to support a precipitation modelling approach was carried out. The modelling approach was applied to the case of calcite precipitation. The model was designed keeping in mind the adaptability to *Generalized Physicochemical Modelling Framework* (Batstone *et al.*, 2012). Furthermore, the authors discussed possible usable control strategies and nutrient recovery models. pH titration tests and constant composition experiments were carried out to define the base line of the model approach and then deduce the environmental factors on the baseline, respectively. Ion-pairing and acid-base reactions were formulated algebraically following the mass action principle. Parameters from thermodynamic databases were used to formulate correct stoichiometry and to assign appropriate equilibrium constants to the chemical reactions. Chemical activities of each component were also considered by multiplying for a correction factor. Davies equation with temperature correction was used to derive the activity coefficients. The pH was predicted by proton balance method, which considers calculations of total hydrogen using a Tableau method. The resulting algebraic equations were solved iteratively by the Newton-Raphson method. Further, precipitation and dissolution of the chemical constituents were considered based on the Saturation Index S_I approach. This approach necessitates the use of predefined conditions based on supersaturation to select or omit a given chemical precipitation reaction. For example, formation of $CaCO_3$ can be considered in the model if the saturation index reaches a value which supports formation of precipitate. In general, three conditions were used: $S_I < 0$, $S_I = 0$ and $S_I > 0$. The semi-empirical precipitation kinetics was based on total dissolved calcium ($S_{Ca^{2+}}$), total dissolved carbonate ($S_{CO_3^{2-}}$) and total particulate calcium carbonate (X_{CaCO_3}). It is important to note that CO_2 stripping was not considered in the model. The reason suggested by the authors corresponds to the low availability of inorganic carbon in liquid phase. This was entrapped by the addition of NaOH which caused little to no stripping of CO_2. The predictions of the aqueous phase model were compared with *Visual MINTEQ* (Version 3.0, Royal Institute of Technology (KTH)). The set of algebraic and ordinary differential equations were solved using an ODE solver and the Newton-Raphson method.

A general methodology to include physicochemical processes in multiphase wastewater treatment models under a plant-wide modelling framework was developed in (Lizarralde *et al.*, 2015). The model emphasized the definition and selection of biochemical, chemical and physicochemical processes taking place in a multiphase environment. The physicochemical plant-wide modelling methodology involved: (1) the definition of the model components and transformations; (2) the mass-transport definition for a multiphase model of wastewater treatment plant; and (3) the numerical solution procedure. The first step in this methodology, demands the compilation of the biochemical, chemical and physicochemical processes. This was achieved by denominating and specifying the relation between: (1) COD, N, P or S removal to biochemical model; (2) acid-base equilibrium and ion-pairing equilibrium to chemical model; and (3) liquid-gas transfer and liquid-solid transfer to physicochemical model. Subsequently, important transformations were individuated and introduced in the model by means of relations between dynamic state variables. The liquid-solid transfer processes involved precipitation and dissolution processes. The major precipitates were defined and included in the model. They include $CaCO_3$, $MgCO_3$, $Ca_3(PO_4)_2$, struvite, k-struvite and newberyite. A kinetic model based on ion activity product (IAP) was reformulated to take into account the spontaneous nucleation in addition to the development of supersaturation, nucleation and growth of the solids. Interestingly, dissolution was considered as the reverse of precipitation process, which made it easier to implement. The ion-pairing processes in the chemical model are built at the end in conjugation with acid-base equilibrium processes. The relevant chemical components and species were selected and represented as generic dissociation reactions. The chemical equilibrium could be solved through ordinary differential equations or by algebraic equations.

5.3.1.2 ADM1-based precipitation models

Soon after the publication of ADM1, the model received a considerable amount of criticism for some errors and omissions (Batstone *et al.*, 2006; Kleerebezem & van Loosdrecht, 2006). These concern mostly the complexity of the model structure, the stoichiometry and inhibition kinetics of the biological processes, the number and the uncertainty of the parameter values. Requests were also received for the inclusion of important anaerobic processes such as sulfate reduction (Batstone *et al.*, 2006; Kleerebezem & van Loosdrecht, 2006), phosphorus metabolism and physicochemical processes (mineral precipitation in particular). These omissions were acknowledged by the authors of the model and a '*Generalized Physicochemical Modelling Framework*' was developed (Mbamba *et al.*, 2015a). However, multiple studies have been carried out with modified versions of ADM1. These studies include extensions for both biochemical (Federovich *et al.*, 2003; Galí *et al.*, 2009; Parker & Wu, 2006; Peiris *et al.*, 2006) and physicochemical processes (Batstone & Keller, 2003; Flores-Alsina *et al.*, 2016; Maharaj *et al.*, 2018; Zhang *et al.*, 2015). Here we discuss the ADM1 extensions to include precipitation processes.

Calcium carbonate ($CaCO_3$) precipitation was implemented as a minor structural modification to the standard ADM1 (Batstone & Keller, 2003). To predict $CaCO_3$ precipitation, three additional state variables ($S_{CO_3^{2-}}$, S_{Ca} and S_{CaCO_3}) representing the concentrations of CO_3^{2-}, Ca^{2+} and $CaCO_3$ respectively, were defined and added. Simple first-order kinetic rate equations for HCO_3^-/CO_3^{2-} acid-base reactions and $CaCO_3$ precipitation reaction were also added. ΔH° values of precipitation and acid base reactions were used to calculate the change in value of the equilibrium coefficients due to temperature. The model was used to assess two case studies: (1) recycled paper-mill wastewater fed to UASB, and (2) gelatine production wastewater fed to solids digester.

A more detailed physicochemical framework was implemented by (Zhang et al., 2015) to simulate the dynamics of calcium caonate ($CaCO_3$), magnesium carbonate ($MgCO_3$), struvite ($MgNH_4PO_3$), magnesium phosphate ($MgHPO_4$) and tricalcium phosphate ($Ca_3(PO_4)_2$). ADM1 was modified by improving the biochemical framework and integrating a more detailed physicochemical (gas transfer and precipitation) framework. The changes in inorganic carbon and nitrogen for decay processes were considered and carbon and nitrogen balances were closed by adding balance terms of inorganic carbon and nitrogen for the microbial decay processes. Two balance terms based on a previous study (Blumensaat & Keller, 2005) were introduced into the original ADM1 to resolve the discrepancies between carbon and nitrogen contents in the degraders and substrates. In the physicochemical framework, an additional gas transfer process was incorporated to account for NH_3. Further, inorganic components for acid-base reactions and solid precipitation were added. The inorganic carbon components included dissolved carbon dioxide, bicarbonate and carbonate. Similarly, inorganic phosphate components such as phosphate, hydrogen phosphate, dihydrogen phosphate and phosphoric acid were introduced into the model. Consequently, the charge balance was modified to accommodate the new components. A second-order irreversible precipitation model was used to improve the ability of the model to simulate non-biologically mediated processes. The model equation used in this study is based on the fundamental relation for crystallization process which can describe precipitation processes better than any simple first-order rate equations based on pseudo-equilibrium. The model was validated with literature data. The model was used to assess the effect of calcium ions, magnesium ions, inorganic phosphorus and inorganic nitrogen on a batch anaerobic digester.

A series of model extensions, aimed at improving the interactions of phosphorus, sulfur and iron, mineral precipitation processes, were implemented in the framework of ADM1. The extended ADM1 was the part of a larger plant-wide model. The model extension (A_3) included: (1) aqueous chemistry model (Solon et al., 2015), and (2) a multiple mineral precipitation model, MMP (Mbamba et al., 2015b). The first submodule estimates the pH by considering ionic behaviour of non-ideality (ion-pairing and ion activity) instead of molar ion concentration. A set of non-linear algebraic equations were used to resolve the aqueous chemistry module. The method includes: (1) Davis equation based ionic strength correction; (2) ion-pairing equilibrium reactions for inorganic carbon, inorganic nitrogen and VFA; and, (3) multi-dimensional Newton-Raphson method to solve the set of algebraic equations. In the second submodule of MMP, a reversible mechanism of precipitation has been implemented using saturation index. This methodology of predicting precipitation reactions has been already explained in detail (Mbamba *et al.*, 2015a, and Section 5.3.1.1). The MMP model simulates the dynamics of calcite (X_{CaCO_3}), agronite ($X_{CaCO_{3a}}$), amorphous calcium phosphate ($X_{Ca_3(PO_4)_2}$), hydroxylapatite ($X_{Ca_5(PO_4)_3(OH)}$), octacalium phosphate ($X_{Ca_8H_2(PO_4)_6}$), struvite ($X_{MgNH_4PO_4}$), newberyte (X_{MgHPO_4}), magnesite (X_{MgCO_3}), k-struvite (X_{KMgPO_4}), iron sulfide ($X_{FeS.}$), iron phosphate ($X_{FePO_4}/X_{Fe_3(PO_4)_2}$), and aluminium phosphate (X_{AlPO_4}).

Recently, a model based on ADM1 has been proposed which explicitly predicts the dynamics of TEs (Fe, Co and Ni) in terms of mineral precipitation in an anaerobic batch reactor (Maharaj *et al.*, 2018). In the biochemical module, the microbial uptake of TE and the TE inhibition on microbial activities have been introduced. A dose response function has been implemented to model the effect of TEs on anaerobic production of methane. The function considers deficiency, activation, inhibition and toxicity of TEs on the microbial groups. In the physicochemical module, ion association/dissociation reactions and liquid-gas transfer reactions are considered for liquid-liquid processes and liquid-gas processes, respectively. Equilibrium reactions are modeled by a set of differential equations. The incorporation of the precipitation reactions in the physicochemical module leads to the definition of new inorganic components in the ADM1 framework. The model takes into consideration inorganic carbonate (e.g. CO_2, HCO_3^- and CO_3^{2-}), inorganic phosphate (e.g. PO_4^{3-}, HPO_4^{2-}, $H_2PO_4^-$ and H_3PO_4) and inorganic sulfide (e.g. HS^- and S^{2-}) components in liquid phase. The components of the three chemical systems (carbonate, phosphate and sulfide) react, in the liquid phase, to form precipitates (e.g. $CoCO_3$, $Co_3(PO_4)_2$, FeS, FeS, $FeCO_3$, $Fe_3(PO_4)_2$, $NiCO_3$, $Ni_3(PO_4)_2$, NiS, $MgNH_4PO_4$), whose formation is governed by the K_{sp} values. A full kinetic framework has been used to implement the precipitation process. The model was implemented on an original code in MATLAB platform and has been solved using 'ode15s' routine. The model was used to study the effect of changes in the initial concentration of sulfur and phosphorous, the effect of nutrient starvation on methane production and the dynamics of TEs in AD.

5.3.2 Adsorption

Adsorption of TEs in AD can proceed through three different mechanisms: (i) adsorption of TEs on precipitates forming in the digester; (ii) complexation of TEs on organic matter released by the microorganisms; and (iii) adsorption of TEs on the surface of the microorganisms. In the first case, when a precipitate is formed it provides many adsorption sites for various cations and anions. In the second and third case, various carboxylic and amino groups of organic surfaces provide sites for the surface complexation of TEs. In general, TE adsorption in biogeochemical systems has been studied and reviewed to a great extent (Konhauser, 2007; Warren & Haack, 2001). However, only few attempts to model the adsorption process in wastewater treatment systems exist in literature (Hauduc *et al.*, 2015), and up to date they have not been implemented in the ADM1 framework to study the effect of TEs adsorption on AD performance. In the following section, we report the modelling efforts to simulate the adsorption process in wastewater environments. For a general overview of the models adopted for biosorption the reader can refer to Papirio *et al.* (2017).

The general strategy to model adsorption processes is derived from geochemical reaction modelling techniques. In this approach, a chemical equilibrium problem is formed comprising the major chemical reactions taking place. Further, the equations are solved simultaneously by the Newton-Raphson algorithm. The equilibrium constants of the individual reactions within the system are of major importance because such models consider every physicochemical reaction as being in equilibrium. For example, the mechanisms behind the chemical-mediated phosphorus system was studied in (Smith *et al.*, 2008). The study was aimed at investigating the major factors affecting phosphorus removal and also the effect of various operational parameters. The model took into consideration all the possible surface reactions on hydrous ferric oxide. The basis of the model is that iron and phosphorus share an oxygen atom. The chemical equilibrium problem was divided in two parts for numerical reasons. The first part solves or quantifies the thermodynamically-favored solid precipitates formed ($Fe(OH)_{3(s)}$ and $FePO_{4(s)}$). Once the precipitated solids are estimated, a second calculation step is performed solving a second equilibrium problem where phosphate is allowed to complex with active sites (adsorption sites) on the precipitated hydrous ferric oxide (HFO). As the authors pointed out, phosphate removal should be considered with phosphate precipitation. However, the surface complexation modelling formalism was adopted in this case to remove phosphate from the system in the earlier reaction step (hydrolysis). This was done by kinetically linking the process to the mixing.

In other cases, the kinetic approach to model the adsorption process in a wastewater system has been adopted. In such case, the adsorption of a chemical species is considered to be kinetically controlled. The kinetic approach to define the physicochemical system is easier to implement because in most of the biochemical framework, AD models are kinetically controlled rather than equilibrium controlled. The process of chemical phosphorus removal in wastewater was kinetically modelled by (Hauduc *et al.*, 2015). Such dynamic model serves as a tool to optimize chemical dosing, taking into consideration the effluent phosphorus concentration due to regulatory reasons. The model predicts the stoichiometry and kinetics of precipitation of hydrous ferrous oxides

(HFO), phosphate adsorption and co-precipitation. Thus, chemical equilibrium dissociation, chemical ion pairing, physical mineral precipitation process, chemical surface complexation, and aging of precipitates were considered in the model. Kinetic models used for adsorption processes in wastewater treatment are limited to pseudo-first-order rates (Lagergren type). The adsorption equation expresses the variation of component concentration as function of a driving force which is the difference between the amounts of the component that should be bound at equilibrium. This is calculated based on the equilibrium constants and actual bonded component. For monodentate and bidentate species the order of the reaction is one and two, respectively. The adsorption sites on the flocs are not equally accessible to phosphorus. More accessible sites are available first, which reduces in number with the course of adsorption. Therefore the rate of adsorption decreases. This decrease in kinetic rate due to a decrease in accessibility of remaining sites have been modeled by multiplying the ratio $(SiteF/SiteT)^n$ to the overall rate, where, n is 1 for monodentate and 2 for bidentate species. The ratio is the amount of free sites (SiteF) to the amount of total sites (SiteT) with a value between 0 and 1.

A biogeochemical framework (CCBATCH) was implemented and expanded by adding surface complexation reactions to the already developed biodegradation and chemical equilibrium sub models (Schwarz & Rittmann, 2007). The surface complexation sub-model included both electrostatic and non-electrostatic surface complexation reactions. Metal complexation by active cells, inert biomass (solids) was described by the electrostatic surface complexation model. The non-electrostatic sub-model was used to describe metal complexation reactions of extracellular polymeric substances (EPS) and biomass-associated-product (BAP). The model assumed a negative charge on the surface of the cell which is due to the protonation/deprotonation of carboxyl, phosphoryl and hydroxyl groups. The interaction of these negatively charged surface species with the adsorbing protons was quantified according to the capacitance model of microbial surface. Further, to understand the biological and non-biological ligand interaction with cell surface leading to metal detoxification a three-step procedure was adopted. Stoichiometry and kinetics of production of reactive ligands by the microbes were mentioned in the first step. In the second step, reactivity of the ligands towards metal was described which indirectly described the speciation of the metal in the system. Lastly, expanded CCBATCH was used to carry out titration reactions simulations. The biogeochemical analyses emphasized the fact that over various organic and inorganic microbial products, sulfide is a major promoter for Zn detoxification. It was shown that, in presence of sulfide, detoxification of Zn occurs while in absence of sulfide, Zn complexes with biogenic organic ligands providing a mechanism for detoxification.

A surface-complexation modelling approach to describe the metal ion interactions with microbial surface has also been studied in (Daughney & Fein, 1998). This modelling technique makes use of stability constants to model metal-microbial surface interaction. However, stability constants for metal-microbial surface complexes, which form the backbone of this modelling technique, are rarely reported because of the intricacies of the experimental methods involved in their determination. Therefore, the stability constants were predicted using a linear free energy approach. In this regard, measured microbial metal-carboxyl stability constants are related to stability constants for aqueous metal-organic acid anion complexes that involve the same metal cation. If the correlation between these two types of stability constants is valid, an unknown value for a metal-carboxyl microbial surface stability constant can be estimated provided the stability constant for the metal-organic acid anion aqueous complex is known. Whether such a relationship could be constrained for a widely studied organic acid anion, it would provide a means of estimating the microbial adsorption behavior of a wide range of metal cations.

5.3.3 Aqueous complexation

Anaerobic digestion is rich in organic matter. The organic matter is composed of biomass, humic substances and other organic molecules from the biodegradation, such as VFAs, alcohol and acetate. Their formation is one of the major biochemical reactions in AD. These organic substrates and humic substances act as chelating agents affecting metal speciation. Thiol-containing organic matter, as represented by cysteine, has also been reported to take part in aqueous complexation with TEs as a strong functional group (Yekta *et al.*, 2014a, b). Additionally, in the course of AD, synthetic chelating agents are occasionally added to improve the digester performance. Synthetic agents, such as EDTA, influence (imparting both positive and negative effects) the bioavailability of TEs in AD by forming stable soluble complexes. Thus, three groups of organic species should be taken into consideration for organic complexation modelling: (1) organic substrates in AD (for example, VFAs); (2) humic substances; and (3) synthetic chelating agents (EDTA/NTA/EDDS). Here we discuss the general methodology developed in (Willet & Rittmann, 2003) and adopted in (Maharaj *et al.*, 2019) to model TEs complexation in the ADM1 framework.

The formation and dissociation of TE complexes follows a kinetically controlled approach. The complexation reaction involving organic ligands and TEs was used here to illustrate the kinetic framework to model complexation reactions. New state variables representing the organic ligand acid/base system and complexes have to be taken into account to predict the pH of the system. The general mechanism of complexation reactions and modelling can be written as follows:

$$Me^x + L^y \xrightarrow{k_1} [MeL]^{x+y} \tag{5.20}$$

$$[MeL]^{x+y} \xrightarrow{k_{-1}} Me^x + L^y \tag{5.21}$$

where k_1 is the $[MeL]^{x+y}$ formation rate constant in $[ML^{-3} T^{-1}]$; k_{-1} is the dissociation constant in $[T^{-1}]$. Equations (5.16) and (5.17) can be rewritten as:

$$Me^x + L^y \overset{K_{[MeL]^{x+y}}}{\longleftrightarrow} [MeL]^{x+y} \qquad (5.22)$$

where $K_{[MeL]^{x+y}}$ is the equilibrium constant for $[MeL]^{x+y}$ in $[ML^{-3}]$. The rate of change of complex over time may be written as:

$$\frac{d[MeL]^{x+y}}{dt} = k_1 [Me^x][L^y] - k_{-1}[MeL]^{x+y} \qquad (5.23)$$

where $[Me^x]$, $[L^y]$, $[MeL]^{x+y}$ are the dynamic state variables for the free TE concentration, organic ligand and TE-organic complex respectively. The formation rate constant, k_1, for all the species and the dissociation rate constant, k_{-1}, can be calculated from the stability constant by a simple relation as:

$$K_{[MeL]^{x+y}} = \frac{k_1}{k_{-1}} \qquad (5.24)$$

5.4 MODELLING THE EFFECT OF TEs ON BIOKINETICS
5.4.1 Modelling the biouptake of TEs

It is widely accepted that metal geochemistry and environmental conditions in an engineered system affect the uptake of TEs by microorganisms (Hudson, 1998; Jansen, 2004; Sunda & Huntsman, 1998; Worms et al., 2006). The bioavailability of metals in the case of AD has been recently studied (Gustavsson, 2012; Jansen, 2004; Yekta et al., 2016). Bioavailability is thought to depend on: (i) the internalization pathway; (ii) specificity of metals to the transporters; (iii) physicochemistry of the bulk phase; (iv) size and nature of microorganisms; (v) concentration of metal; and (vi) metal speciation.

To effectively affect the biochemical systems, the required metals have to be transported from the bulk solution into the Cytosol (Figure 5.3). In this process metals diffuse through the external medium to the surface of the organism, they are subsequently adsorbed to the microbial surface and later internalized by the corresponding mechanisms. These transport processes control the overall metal bio-uptake rates (Jansen, 2004; Koster & Leeuwen, 2004; Worms et al., 2006). The binding of metals to the surface of the microorganisms are thought to be carried out by the extracellular polysaccharides (Bhaskar & Bhosle, 2006; d'Abzac et al., 2010; Malik, 2004; Loaec et al., 1997). Therefore, the rate of extracellular polysaccharide formation on the microorganisms might be linked to the presence of a certain amount of a particular metal in the surrounding environment (Aquino & Stuckey, 2004). The surface-bound metals are destined to be internalized, accumulate on the surface or dissociate back to the bulk phase (Worms et al., 2006). The internalization of metals depends largely on the transport system (Hudson, 1998; Jansen, 2004; Worms et al., 2006). The

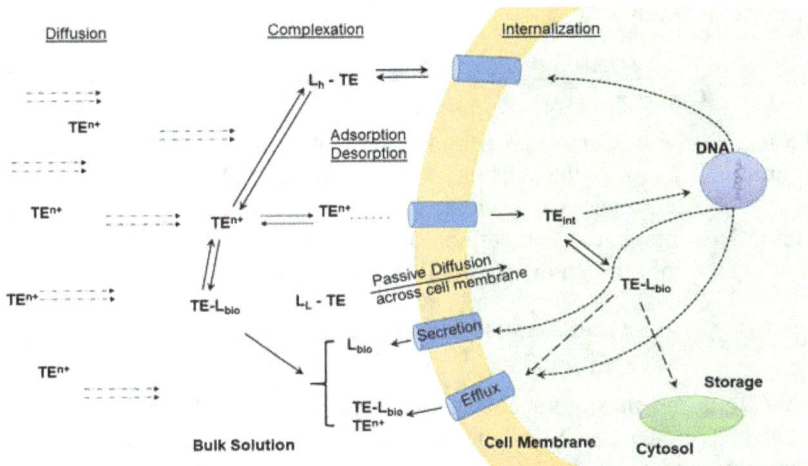

Figure 5.3 Various physico-chemical and transport mechanisms affecting TE internalization in a microbial cell. L and TE^{n+} refer to ligand and free trace element respectively. Subscripts int: internalized; bio: microbial; h: hydrophilic; L: lipophilic (adapted from Worms *et al.*, 2006).

hydrophobic nature of biological membranes limits the movement of molecules by passive diffusion. Only neutral and non-polar molecules move across the membrane by passive diffusion. Difference in concentration gradient across the membrane is the driving force for passive diffusion of molecules across the membrane. Metal species in AD are generally hydrophilic in nature and their transport is mediated by specific metal-binding proteins (Worms *et al.*, 2006). For example, metals are transported as metal ions involving ATPases, natural resistance associated microphage proteins and zinc-regulated or iron-regulated transporter. These two steps of metal binding to the surface and metal internalization by specific transporters have been expressed by Michaelis-Menten kinetics (Jansen, 2004). Further, the mass balance for total TE has been established. Starting with the case of growth limitation, metal concentration in the bulk phase can be related to growth by Monod kinetics. As Monod kinetics assumes the external metal concentration in the bulk phase, it might become erroneous to model internal metal concentration. Droop model of relating internal metal concentration to the growth renders an option here (Droop, 1983; Jansen, 2004). Microbial growth rate was defined as:

$$\frac{dx}{dt} = \mu x \tag{5.25}$$

where, μ is the growth rate constant, t is the time, and x is the biomass concentration. The limitation of growth rate due to substrate and available metal ions was represented as follows:

$$\mu = \mu_{max} \frac{C_{MeOH}}{C_{MeOH} + K_{MeOH}} \frac{(Q_{Co} - Q_{Co,min})}{Q_{Co}} \frac{(Q_{Ni} - Q_{Ni,min})}{Q_{Ni}} \tag{5.26}$$

where μ_{max} is the maximum growth rate constant, C_{MeOH} is the primary substrate concentration, K_{MeOH} is the substrate Monod constant, Q_{Co} is the cobalt content; $Q_{Co,min}$ is the minimum Co content, Q_{Ni} is the Ni content and $Q_{Ni,min}$ is the minimum Ni content. The competition of metal ions for the same transporter was taken into account through the uptake flux rate as follows:

$$J_{Co,in} = J_{max} \frac{C_{Co^{2+}}/K_{JM,Co^{2+}}}{(C_{Ni^{2+}}/K_{JM,Ni^{2+}}) + (C_{Co^{2+}}/K_{JM,Co^{2+}}) + 1} \tag{5.27}$$

where J_{max} is the maximum uptake flux; $K_{JM,Co^{2+}}$ is the Co Michaelis-Menten constant, $K_{JM,Ni^{2+}}$ is the Ni Michaelis-Menten constant, $C_{Co^{2+}}$ is the free Co^{2+} concertation, and $C_{Ni^{2+}}$ is the free Ni^{2+} concentration. The excretion of metal ions was modelled as a first order process as follows:

$$J_{Co,eff} = k_{eff,Co}(Q_{Co} - Q_{Co,eff,min}) \tag{5.28}$$

where $k_{eff,Co}$ is the efflux rate constant and $Q_{Co,eff,min}$ is a threshold value below which no efflux takes place. The mass balances were established based on the above-mentioned rates. The mass balance for metal content of the cells was represented as follows:

$$\frac{dQ_{Co}}{dt} = J_{Co,in}(t) - J_{Co,eff}(t) - \mu(t)Q_{Co}(t) \tag{5.29}$$

The amount of total dissolved metal was accounted for with the following mass balance:

$$\frac{dQ_{Co,total}}{dt} = -(J_{Co,in}(t) - J_{Co,eff}(t))x(t) \tag{5.30}$$

5.4.2 Dose-response modelling of TE effect in AD

Elements in trace amounts such as Fe, Zn, Ni, Co, Mo, Cu, Mn, Se, Mo, W, and B considerably affect biogas production in anaerobic digesters by stimulating and enhancing microbial activity (Choong et al., 2016; Federation, 2014; Feng et al., 2010; Fermoso et al., 2009; Oleszkiewicz & Sharma, 1990; Romero-güiza et al., 2016). Optimal levels of TEs should be maintained by adding externally the appropriate micronutrients, especially when the feedstock used for AD is deficient in these elements. It is a common industrial practice to supply these elements externally to ensure that these will be in bioavailable form to cover the nutritional needs of the microorganisms in the anaerobic environment. However, increased concentrations of the same elements can act as inhibiting or toxic factor and result in a reduced biogas production or complete reactor failure (Feng *et al.*, 2010; Hickey *et al.*, 1989; Lin, 1992). Working within the optimum concentration range is required to secure balancing between deficiency and toxicity and ensure minimum operating cost for additives supply. This behaviour is conceptually presented in Figure 5.4(a) for an essential TE. When an element is considered 'non-essential' for microbial growth, the deficiency part of Figure 5.4(a) is narrowed or supressed. In such cases, the effect of the element on the microbial growth presented conceptually in Figure 5.4(b) can be observed. Mathematically, this behaviour is approached by dose-response models. These models are extensively used in pharmacokinetic (PK) studies (Zhao & Yang, 2014), where the effect of dosing a medicine to a sample population is monitored by measuring some physiological response. The same concept can be applied for modelling the effect of TEs on the biological activity in anaerobic digesters.

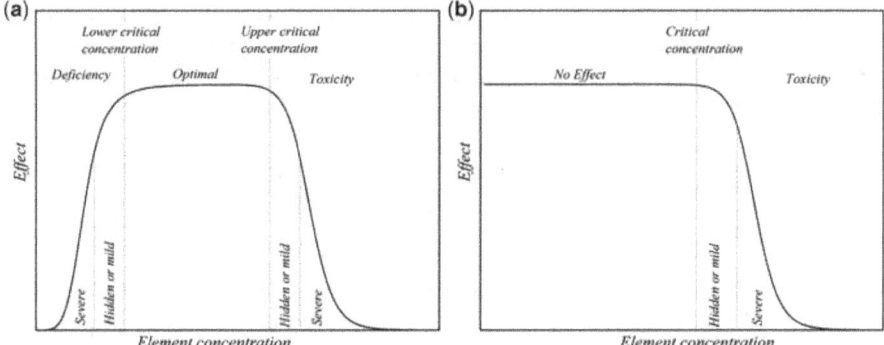

Figure 5.4 A conceptual dose-response curve for (a) an essential and (b) a non-essential element.

In typical dose response studies only the part of the response up to the maximum effect level is modelled (left hand side of Figure 5.4(a)). A sigmoidal curve usually adequately describes this part of the curve. A number of mathematical expressions simulate such a sigmoidal response. Selecting the appropriate model depends on the following aspects.

Theoretical basis of the model: Very few sigmoidal models are derived from a theoretical analysis of the dose-response effect. The majority of the models are empirical equations or mathematical expressions sharing sigmoidal curve characteristics.

Number of parameters in the model: The number of parameters affects the shape (steepness) and position of the reflection point of the sigmoidal curve. It usually ranges from two to five (excluding y_0 and y_{max} values). Equations with fewer parameters are preferred for simplicity. For the case of simple models with only two parameters, the linearized form of the equation can be used to determine the best-fit values simulating an experimental set of data. Generally this procedure is carried out by nonlinear regression analysis.

Position of the reflection point: The position of reflection point of the sigmoidal curve is also a significant criterion to select a dose response model for simulation. The reflection point can be either fixed positioned at 50% of effective dose (ED_{50}) (e.g. E_{max} model) or parametrically adjustably above or below ED_{50}.

The concept of dose-response modelling can be directly applied in modelling the effect of TEs in anaerobic digesters. The effective concentration applied for each TE should be mathematically correlated with the induced response in the system. Therefore two critical parameters should be defined: (a) which is the selected response and at which time should the system be sampled (observed y values), and (b) which effective concentration of the TE produces a certain level of response (measured x values). Ideally, direct effects such as increased microbial activity related with higher observed growth rates and/or increased yield coefficients should be quantified experimentally. Microbial growth rate expressions can be modified appropriately by including terms for the effect of TEs. When Monod kinetic expressions are used, these can be modified as follows:

$$r_i = \mu_{max,i} \frac{S_i}{K_i + S_i} \cdots \frac{S_j}{K_j + S_j} \dots f(I_i) \dots f(I_j) \dots f(pH) f(T) \dots f(TE_i) \dots f(TE_j)$$

(5.31)

where r_i is the specific growth rate of microorganisms in process i [ML^{-3} T^{-1}]; $\mu_{max,i}$ is the maximum specific growth rate of microorganisms in process i [T^{-1}]; $S_i \dots S_j$ is the concentration of limiting substrates $i \dots j$ e.g. carbon source, nitrogen source etc. [ML^{-3}]; $K_i \dots K_j$ is the half saturation constant for the components $i \dots j$ [ML^{-3}]; $f(I_i) \dots f(I_j)$ are the factors for inhibiting substances e. g. free ammonia, butyrate (dimensionless); $f(pH)$ is the effect of pH on the specific growth rate of microorganisms in process I; $f(T)$ is the effect of temperature on the specific growth rate of microorganisms in process i (dimensionless); $f(TE_i) \dots f(TE_j)$ is the effect of TEs $i \dots j$ on the specific growth rate of microorganisms in process i (dimensionless).

In a traditional modelling approach biomass yield (Y) and related stoichiometric coefficients are considered constant, although they may vary significantly during the growth stages of the microbial biomass and the physicochemical conditions in cell's environment (e.g. TEs). When the yield is affected by the presence of essential TEs then a similar approach can be used:

$$Y = Y_0 f(TE_i) \dots f(TE_j)$$

(5.32)

where Y is the observed yield coefficient in the presence of TEs; Y_0, is the maximum yield coefficient at the optimum level of TEs; $f(TE_i) \dots f(TE_j)$ is the dose response equations related with the TEs $i \dots j$.

Assuming that an element is essential for growth, any sigmoidal curve describing the left hand side of Figure 5.4(a) can be used as a modifying function. The model would result in non-growth conditions when the element is absent. This corresponds to complete reactor failure due to severe element deficiency. Conditions of mild deficiency are also modeled by equations (5.27) or (5.28) at moderate element concentration. Finally, saturation is predicted when the concentration of TEs are within the optimal range. Equations (5.27) and (5.28) should be modified appropriately when the observed rates and/or yields are not bound between 0 and 1 as in a 'typical' sigmoidal curve. For example when the baseline effect is different than zero the appropriate $f(TE_j)$ functions should be used. Toxicity modelling is also possible by selecting appropriate mathematical expression such as equations in Table 5.5.

Estimating the bio-kinetic parameters for an anaerobic digester, by fitting a model to experimental data, assumes that the only limiting or inhibiting substances that affect the process are these already included in the kinetic expressions of the model (e.g. first part of equation 5.27). However under severe or mild TEs deficiency (correspondingly toxicity) the bio-kinetic estimated parameters might be significantly misleading as the system does not perform under optimum conditions. The use of an updated revised model would result in a different set of parameters that include also the effect of the trace elements in the system.

Another significant aspect which should also be addressed is the selection of the classes of the microbial consortia present in an anaerobic digester which are mostly affected by the elements deficiency. It has been shown that the microorganisms that are more sensitive to TEs fluctuation and deficiency are those belonging to the group of methanogenic archaea (Glass & Orphan, 2012). Therefore, the critical step in this modelling approach is to focus primarily on the metabolic activity of this group by modifying the appropriate kinetic expression and/or yield coefficients.

It is also important to define what is considered the effective concentration (dose) of a TE in an anaerobic bioreactor resulting to a certain level of stimulating response (or correspondingly toxicity effect). In pharmacokinetic studies when a new medicine is tested in a sample population, a prescribed amount is administered, while the resulted physiological response is monitored (e.g. changes in blood pressure). Even if the drug is distributed differently in the human organs, excreted, accumulated or metabolized, the total amount of the drug administered is considered as the effective dose. However, in those studies factors such as age, sex, race and so on, may significantly differentiate the results. Similarly, in the case of anaerobic digesters, the total concentration of trace elements could be used as a rough indication of the dose to be used in dose-response equations. However, it is realized that a significant fraction of the total element in a bioreactor is in non-readily bioavailable form and will not affect directly biogas production (Thanh et al., 2015). Part of the element can be distributed according to the

Table 5.5 List of dose-response functions.

No.	Name	Dose-response Function	General Trend
1	Hill equation	$y = y_{max} \dfrac{x^n}{a^n + x^n}$ where • y_{max}, a are parameters of the model • n is the Hill coefficient	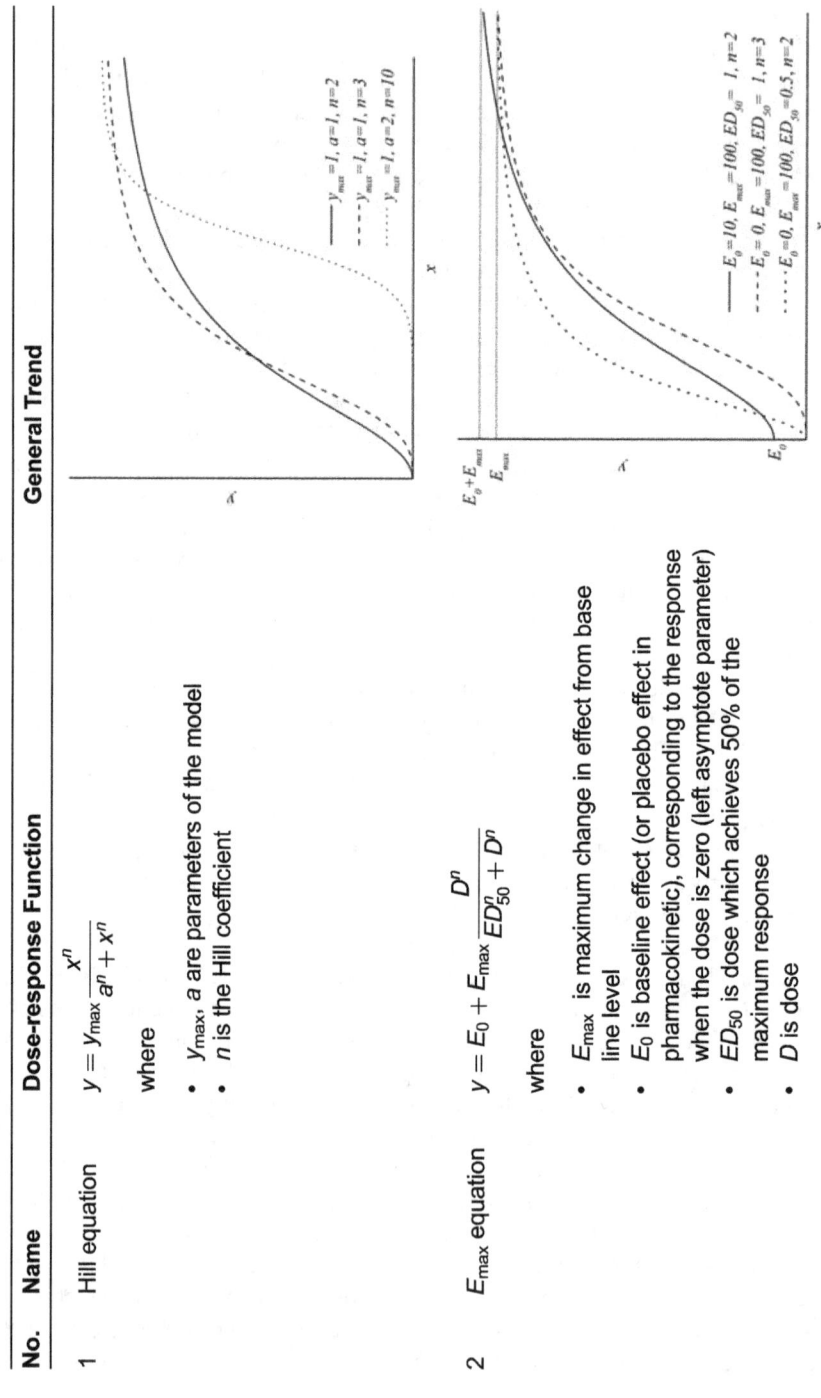
2	E_{max} equation	$y = E_0 + E_{max} \dfrac{D^n}{ED_{50}^n + D^n}$ where • E_{max} is maximum change in effect from base line level • E_0 is baseline effect (or placebo effect in pharmacokinetic), corresponding to the response when the dose is zero (left asymptote parameter) • ED_{50} is dose which achieves 50% of the maximum response • D is dose	

3   Logarithmic reciprocal equation

$y = e^{a-b/x}$

where

- a and b are parameters of the model

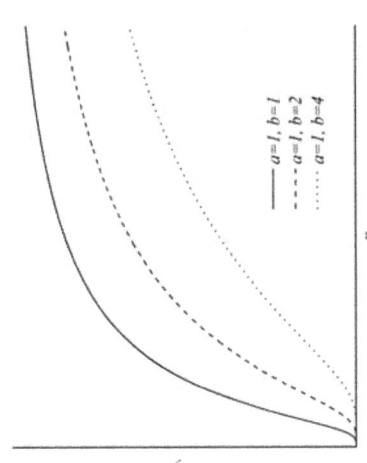

a=1, b=1
a=1, b=2
a=1, b=4

4   Logistic equation

$$y = \frac{y_{max}}{1 + e^{-(a+bx)}} = \frac{y_{max}}{1 + a'e^{-bx}}$$

where

- y_{max}, a or a' and b are parameters of the model

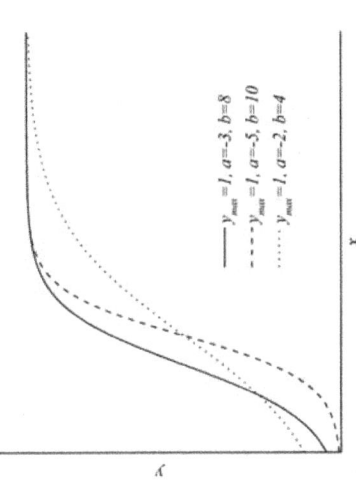

$y_{max} = 1, a = -3, b = 8$
$y_{max} = 1, a = -5, b = 10$
$y_{max} = 1, a = -2, b = 4$

(Continued)

Table 5.5 List of dose-response functions (*Continued*).

No.	Name	Dose-response Function	General Trend
5	Log-Logistic equation	$y = \dfrac{y_{max}}{1 + ae^{-b\,ln(x)}}$ where • y_{max}, a and b are parameters of the model	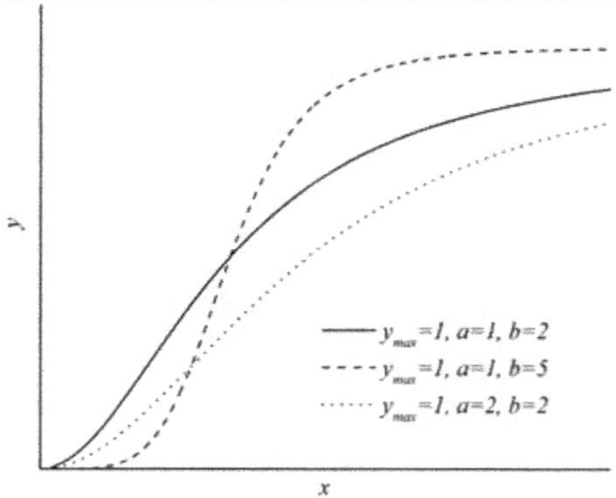
6	Gompertz equation	$y = y_{max}e^{-e^{-(a+bx)}} = y_{max}e^{-a\,e^{-bx}}$ where • y_{max}, a and b are parameters of the model.	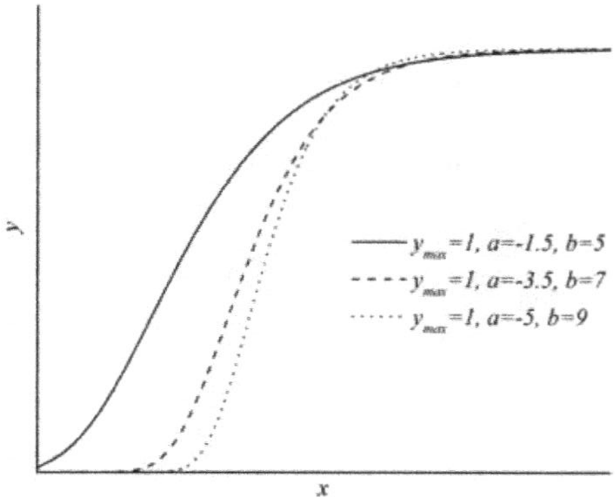

Legend for No. 5 curves:
— $y_{max}=1, a=1, b=2$
- - - $y_{max}=1, a=1, b=5$
······ $y_{max}=1, a=2, b=2$

Legend for No. 6 curves:
— $y_{max}=1, a=1.5, b=5$
- - - $y_{max}=1, a=3.5, b=7$
······ $y_{max}=1, a=5, b=9$

7 | Weibull equation

$$y = y_{max} - ae^{-bx^c}$$

where

- y_{max}, a, b and c are parameters of the model

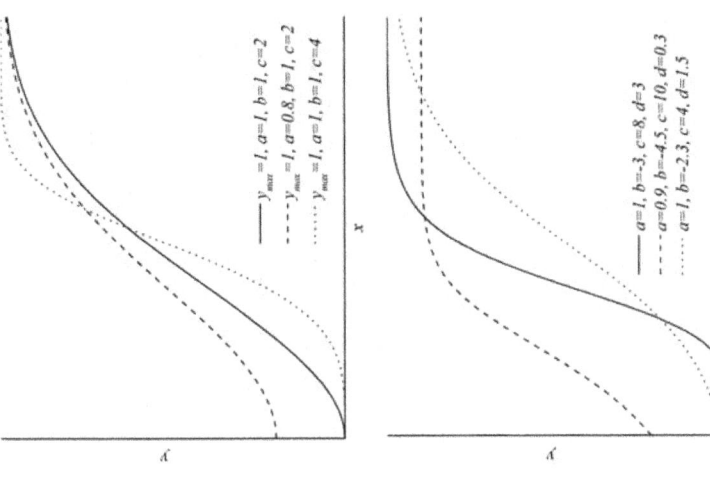

$$\text{—— } y_{max} = 1, a=1, b=1, c=2$$
$$\text{--- } y_{max} = 1, a=0.8, b=1, c=2$$
$$\text{········ } y_{max} = 1, a=1, b=1, c=4$$

8 | Richard's equation

$$y = \frac{a}{(1 + e^{-(b+cx)})^d}$$

where

- a, b, c and d are parameters of the model

$$\text{—— } a=1, b=-3, c=8, d=3$$
$$\text{--- } a=0.9, b=-4.5, c=10, d=0.3$$
$$\text{········ } a=1, b=-2.3, c=4, d=1.5$$

(*Continued*)

Table 5.5 List of dose-response functions (*Continued*).

No.	Name	Dose-response Function	General Trend
9	Equations simulating stimulation and toxicity effects	$$y = y_0 + \frac{y_{max} - y_0}{\left(1 + \dfrac{EC_{50}}{x}\right)^a \left(1 + \dfrac{x}{IC_{50}}\right)^b}$$ where • y is the recorded response • y_0 is the baseline of the recorded response at concentration levels approaching zero • y_{max} is the maximum observed response • EC_{50} is the concentration of trace element where 50% of the maximum effect is achieved • IC_{50} is the concentration of the element where 50% inhibition (toxicity) is observed • a and b are parameters that affect the slope of the left hand side and right hand side of the curve respectively	$y_0=0$, $y_{max}=1$, $EC_{50}=100$, $IC_{50}=500$, $a=2$, $b=10$; $y_0=0.1$, $y_{max}=1.3$, $EC_{50}=100$, $IC_{50}=500$, $a=2$, $b=10$; $y_0=0$, $y_{max}=1$, $EC_{50}=100$, $IC_{50}=500$, $a=28$, $b=20$; $y_0=0.1$, $y_{max}=1.1$, $EC_{50}=100$, $IC_{50}=500$, $a=2$, $b=10$
10	Equations simulating hermetic effects	$$y = \frac{(x + a)}{b + c(x + a) + d(x + a)^2}$$ where • a, b, c and d are parameters of the model	$a=0.03$, $b=0.022$, $c=0.138$, $d=4.576$; $a=0.01$, $b=0.022$, $c=0.138$, $d=4.576$; $a=0.03$, $b=0.012$, $c=0.138$, $d=4.576$; $a=0.03$, $b=0.022$, $c=0.258$, $d=4.576$; $a=0.03$, $b=0.022$, $c=0.138$, $d=3.576$

following processes (Callander & Barford, 1983a, b): reversible or irreversible adsorption on the surface of solids (Section 5.3.2); precipitation to inorganic minerals by abiotic reactions (e.g. hydroxide, carbonate, phosphate, sulfide precipitates; Section 5.3.1); volatilization of the elements that produce inorganic or organic volatile compounds (e.g. volatile selenium compounds); redox reactions that affect the solubility of the element (e.g. Cr(III)/Cr(VI), Fe(II)/Fe(III), Se(VI)/Se(IV)/ Se(0)/ Se(-II)); complexation by organic ligands (e.g. proteins, humic acids, fulvic acids, yeast extract, EDTA, NTA, EDDS; Section 5.3.3) (Gonzalez-Gil et al., 2003; Zhang et al., 2015) and species distribution according to the element speciation (e.g. poly hydroxyl species) (Yekta et al., 2016).

Alternatively, as a rough estimate of the effective concentration (dose) it would be appropriate to use the soluble fraction of the TE present in the reactor liquid phase. However even under this convention, metal complexes with organic macromolecules could still be in non-bioavailable form. Thus, it is of primary concern to determine which chemical species of the TE are actually considered directly bioavailable to enhanced biogas production and system stability (Yekta et al., 2016).

From the mathematical point of view, inclusion of terms for TEs in an existing model unavoidably increases model complexity. For each TE a new state variable in the liquid phase should be introduced. In addition, a set of at least two new parameters should be provided to include the appropriate sigmoidal shape equation in the model. The complexity increases considerably when different species of the TE should also be included as well as speciation modelling equations to calculate the concentration of these species. These algebraic equations corresponding to the equilibrium law for each reaction usually are treated as instantaneous or fast dynamic equations. Furthermore, processes such as adsorption, complexation or precipitation of the TEs require definition of additional processes and their corresponding rates or equilibrium equations. Providing coherent initial conditions for such a complex system is crucial for the initialization of the integration algorithm. Additionally, the difference in time scales between slow biological actions and fast dynamic or equilibrium abiotic reactions results in system stiffness. This is the cause of solution failure in the propagation of a solution algorithm.

5.5 CONCLUSION

This chapter reviewed the modelling aspects describing the main processes that affect TE dynamics in AD environment. The consolidated knowledge from the literature indicates that the effect and fate of TEs in AD system can be modelled to predict the dynamic behavior of TEs in AD. However, there is a need to improve the currently available AD models to account for precipitation, adsorption, aquatic complexation and bio-uptake of TE, so as to quantify their fate and effect. While problems may arise due to increase in model complexity, the modelling approach should be compatible with the existing AD framework (ADM1). Consequently, taking into account the multiplicity of processes that affect TE dynamics in AD systems as well as the importance of these processes in defining the biology and physicochemistry of the AD system, the need for a TE module embedded within the ADM1 model has been recognized. This sets the priorities in the efforts towards 'mathematical model-based controlled TE dosing' in engineered AD systems.

REFERENCES

Ali M. I. and Schneider P. A. (2008). An approach of estimating struvite growth kinetic incorporating thermodynamic and solution chemistry, kinetic and process description. *Chemical Engineering Science*, **63**(13), 3514–3525.

Allison J. D., Brown D. S. and Novo-Gradac K. J. (1991). MINTEQA2/PRODEFA2, a Geochemical Assessment Model for Environmental Systems: Version 3.0 User's Manual. U. S. Environ. Prot. Agency EPA/600/3-91/021 115. USEPA, Washington, DC.

Aquino S. F. and Stuckey D. C. (2004). Soluble microbial products formation in anaerobic chemostats in the presence of toxic compounds. *Water Research*, **38**, 255–266, https://doi.org/10.1016/j.watres.2003.09.031

Banerjee R. and Ragsdale S. W. (2003). The many faces of vitamin B12. *Annual Review of Biochemistry*, **72**, 209–247, https://doi.org/10.1146/annurev.biochem.72.121801.161828

Barat R., Montoya T., Seco A. and Ferrer J. (2011). Modelling biological and chemically induced precipitation of calcium phosphate in enhanced biological phosphorus removal systems. *Water Research*, **45**, 3744–3752, https://doi.org/10.1016/j.watres.2011.04.028

Barrera E. L., Spanjers H., Dewulf J., Romero O. and Rosa E. (2013). The sulfur chain in biogas production from sulfate-rich liquid substrates: A review on dynamic modelling with vinasse as model substrate. *Journal of Chemical Technology and Biotechnology*, **88**, 1405–1420, https://doi.org/10.1002/jctb.4071

Barrera E. L., Spanjers H., Solon K., Amerlinck Y., Nopens I. and Dewulf J. (2015). Modelling the anaerobic digestion of cane-molasses vinasse: Extension of the Anaerobic Digestion Model No. 1 (ADM1) with sulfate reduction for a very high strength and sulfate rich wastewater. *Water Research*, **71**, 42–54, https://doi.org/10.1016/j.watres.2014.12.026

Batstone D. J. and Keller J. (2003). Industrial applications of the IWA anaerobic digestion model No.1 (ADM1). *Water Science & Technology*, **47**, 199–206.

Batstone D. J., Keller J., Angelidaki I., Kalyuzhnyi S. V., Pavlostathis S. G., Rozzi A., Sanders W. T., Siegrist H. and Vavilin V. A. (2002). The IWA Anaerobic Digestion Model No 1 (ADM1). *Water Science & Technology*, **45**, 65–73, https://doi.org/10.2166/wst.2008.678

Batstone D. J., Keller J. and Steyer J. P. (2006). A review of ADM1 extensions, applications, and analysis: 2002–2005. *Water Science & Technology*, **54**, 1–10, https://doi.org/10.2166/wst.2006.520

Batstone D. J., Amerlinck Y., Ekama G., Goel R., Grau P., Johnson B., Kaya I., Steyer J. P., Tait S., Takacs I., Vanrolleghem P. A., Brouckaert C. J. and Volcke E. (2012). Towards a generalized physicochemical framework. *Water Science & Technology*, **66**, 1147–1161, https://doi.org/10.2166/wst.2012.300

Bhaskar P. and Bhosle N. (2006). Bacterial extracellular polymeric substances (EPS): A carrier of heavy metals in the marine food chain. *Environment International*, **32**, 1–8.

Blumensaat F. and Keller J. (2005). Modelling of two-stage anaerobic digestion using the IWA Anaerobic Digestion Model No. 1 (ADM1). *Water Research*, **39**, 171–83, https://doi.org/10.1016/j.watres.2004.07.024

Callander I. J. and Barford J. P. (1983a). Precipitation, chelation, and the availability of metals as nutrients in anaerobic digestion. II. Applications. *Biotechnology and Bioengineering*, **25**, 1959–1972, https://doi.org/10.1002/bit.260250806

Callander I. J. and Barford J. P. (1983b). Metals as Nutrients in Anaerobic Digestion. *Biotechnology and Bioengineering*, **25**, 1959–1972.

Chen Y., Cheng J. J. and Creamer K. S. (2008). Inhibition of anaerobic digestion process: A review. *Bioresource Technology*, **99**(10), 4044–4064, https://doi.org/10.1016/j.biortech.2007.01.057.

Choong Y. Y., Ismail N., Abdullah A. Z. and Yhaya M. F. (2016). Impacts of trace element supplementation on the performance of anaerobic digestion process: A critical review. *Bioresource Technology*, **209**, 369–379, https://doi.org/10.1016/j.biortech.2016.03.028

Copp J. B., Jeppsson U. and Rosen C. (2003). Towards an ASM1 − ADM1 State Variable Interface for Plant-Wide Wastewater Treatment Modelling. *Proceeding Water Environmental Federation*, 498–510, https://doi.org/10.2175/193864703784641207

d'Abzac P., Bordas F., van Hullebusch E., Lens P. N. L. and Guibaud G. (2010). Metal binding properties of extracellular polymeric substances extracted from anaerobic granular sludges. *Colloids Surfaces B Biointerfaces*, **80**, 161–168, https://doi.org/10.1016/j.colsurfb.2010.05.043

Daughney C. J. and Fein J. B. (1998). The Effect of Ionic Strength on the Adsorption of H+, Cd2+, Pb2+, and Cu2+ by Bacillus subtilis and Bacillus licheniformis: A Surface Complexation Model. *Journal of Colloid and Interface Science*, **198**, 53–77, https://doi.org/10.1006/jcis.1997.5266

Dezham P., Rosenblum E. and Jenkins D. (1988). Digester gas control using iron salts. *Journal -Water Pollution Control Federation*, **60**(4), 514–517.

Donoso-Bravo A., Bandara W. M. K. R. T. W., Satoh H. and Ruiz-Filippi G. (2013). Explicit temperature-based model for anaerobic digestion: Application in domestic wastewater treatment in a UASB reactor. *Bioresource Technology*, **133**, 437–442, https://doi.org/10.1016/j.biortech.2013.01.174

Droop M. R. (1983). 25 Years of Algal Growth Kinetics A Personal View. *Botanica Marina*, **26**, 99–112.

Ebigbo A., Phillips A., Gerlach R., Helmig R., Cunningham A. B., Class H. and Spangler L. H. (2012). Darcy-scale modeling of microbially induced carbonate mineral precipitation in sand columns. *Water Resources Research*, **48**(7).

England J. L. (2013). Statistical physics of self-replication. *Journal of Chemical Physics*, **139**, 121923, https://doi.org/10.1063/1.4818538

Federation W. E. (2014). Mineral nutrient requirements for high-rate methane fermentation of acetate at low In: SRTM. Takashima and R. E. Speece (eds), *Research Journal of the Water Pollution Control Federation*, **61**(11/12), http://www.jst 61, 1645–1650.

Federovich V., Lens P. and Kalyuzhnyi S. (2003). Extension of Anaerobic Digestion Model No. 1. Appl. *Biochemistry & Biotechnology*, **109**, 33–45, https://doi.org/10.1385/ABAB:109:1-3:33

Feng X. M., Karlsson A., Svensson B. H. and Bertilsson S. (2010). Impact of trace element addition on biogas production from food industrial waste – Linking process to microbial communities. *FEMS Microbiology Ecology*, **74**, 226–240, https://doi.org/10.1111/j.1574-6941.2010.00932.x

Fermoso F. G., Bartacek J., Jansen S. and Lens P. N. L. (2008). Metal supplementation to UASB bioreactors: From cell-metal interactions to full-scale application. *The Science of the Total Environment*, **407**, 3652–3667, https://doi.org/10.1016/j.scitotenv.2008.10.043

Fermoso F. G., Bartacek J., Jansen S. and Lens P. N. L. (2009). Metal supplementation to UASB bioreactors: From cell-metal interactions to full-scale application. *The Science of the Total Environment*, **407**, 3652–3667, https://doi.org/10.1016/j.scitotenv.2008.10.043

Fermoso F. G., Hullebusch E. D., Van Guibaud G., Collins G., Svensson B. H., Vink J. P. M., Esposito G. and Frunzo L. (2015). Fate of Trace Metals in Anaerobic Digestion. Trace metal microbiology. In: Biogas Science and Technology, A. Gübitz, A. Bauer, G. Bochmann, A. Gronauer and S. Weiss (eds), Springer, Berlin, https://doi.org/10.1007/978-3-319-21993-6

Finazzo C., Harmer J., Bauer C., Jaun B., Duin E. C., Mahlert F., Goenrich M., Thauer R. K., Van Doorslaer S. and Schweiger A. (2003). Coenzyme B induced coordination of coenzyme M via its thiol group to Ni(I) of F430 in active methyl-coenzyme M reductase. *Journal of the American Chemical Society*, **125**, 4988–4989, https://doi.org/10.1021/ja0344314

Flores-Alsina X., Solon K., Mbamba C. K., Tait S., Gernaey K. V., Jeppsson U. and Batstone D. J. (2016). Modelling phosphorus (P), sulfur (S) and iron (Fe) interactions for dynamic simulations of anaerobic digestion processes. *Water Research*, **95**, 370–382, https://doi.org/10.1016/j.watres.2016.03.012

Flynn S. L., Szymanowski J. E. and Fein J. B. (2014). Modeling bacterial metal toxicity using a surface complexation approach. *Chemical Geology*, **374**, 110–116.

Galí A., Benabdallah T., Astals S. and Mata-Alvarez J. (2009). Modified version of ADM1 model for agro-waste application. *Bioresource Technology*, **100**, 2783–2790, https://doi.org/10.1016/j.biortech.2008.12.052

Glass J. B. and Orphan V. J. (2012). Trace metal requirements for microbial enzymes involved in the production and consumption of methane and nitrous oxide. *Front Microbiology*, **3**, 1–20, https://doi.org/10.3389/fmicb.2012.00061

Gonzalez-Gil G. and Kleerebezem R. (1999). Effects of Nickel and Cobalt on Kinetics of Methanol Conversion by Methanogenic Sludge as Assessed by On-Line CH 4 Monitoring *Appl. Environmental Biology*, **65**, 1789–1793.

Gonzalez-Gil G., Jansen S., Zandvoort M. H. and Van Leeuwen H. P. (2003). Effect of yeast extract on speciation and bioavailability of nickel and cobalt in anaerobic bioreactors. *Biotechnology and Bioengineering*, **82**, 134–142, https://doi.org/10.1002/bit.10551

Gu Y., Zhang Y. and Zhou X. (2015). Effect of Ca(OH)2pretreatment on extruded rice straw anaerobic digestion. *Bioresource Technology*, **196**, 116–122.

Gustavsson J. (2012). Cobalt and Nickel Bioavailability for Biogas Formation. PhD thesis, University of Linkoping, Sweden.

Hanhoun M., Montastruc L., Azzaro-pantel C., Biscans B., Frèche M. and Pibouleau L. (2011). Temperature impact assessment on struvite solubility product: A thermodynamic modelling approach. *Chemical Engineering Journal*, **167**, 50–58, https://doi.org/10.1016/j.cej.2010.12.001

Hauduc H., Smith S., Szabo A., Murthy S. and Tak I. (2015). A dynamic physicochemical model for chemical phosphorus removal. *Water Research*, **73**, 157–170.

Henze M., Gujer W., Mino T., Matsuo T., Wentzel M. C., Marais G. V. R. and Van Loosdrecht M. C. M. (1999). Activated Sludge Model No.2d, ASM2d. *Water Science & Technology*, **39**, 165–182, https://doi.org/10.1016/S0273-1223(98)00829-4

Hickey R. F., Vanderwielen J. and Switzenbaum M. S. (1989). The effect of heavy metals on methane production and hydrogen and carbon monoxide levels during batch anaerobic sludge digestion. *Water Research*, **23**, 207–218, https://doi.org/10.1016/0043-1354 (89)90045-6

Hille R., Van Foster T., Storey A. and Duncan J. (2004). Heavy metal precipitation by sulphide and bicarbonate: evaluating methods to predict anaerobic digester overflow performance. In: Mine Water 2004–Proceedings Int. Mine Water Assoc. Symp, A. P. Jarvis, B. A. Dudgeon and P. L. Younger (eds), **2**, 141–150.

Huang C., Lai J., Sun X., Li J., Shen J., Han W. and Wang L. (2016). Enhancing anaerobic digestion of waste activated sludge by the combined use of NaOH and Mg(OH)2: Performance evaluation and mechanism study. *Bioresource Technology*, **220**, 601–608.

Hudson R. J. M. (1998). Which aqueous species control the rates of trace metal uptake by aquatic biota? Observations and predictions of non-equilibrium effects. *The Science of the Total Environment*, **219**, 95–115, https://doi.org/10.1016/S0048-9697(98)00230-7

Jansen S. (2004). Speciation and bioavailability of cobalt and nickel in anaerobic wastewater treatment. PhD thesis, Leerstoelgroep Fysische Chemie en Kolloidkunde, Wageningen Univeristy, The Netherlands.

Jeen S. W., Amos R. T. and Blowes D. W. (2012). Modeling gas formation and mineral precipitation in a granular iron column. *Environmental Science & Technology*, **46** (12), 6742–6749.

Jiang B. (2006). The effect of trace elements on the metabolism of methanogenic consortia. PhD thesis, Wageningen University, The Netherlands.

Jimenez J., Latrille E., Harmand J., Robles A., Ferrer J., Gaida D., Wolf C., Mairet F., Bernard O., Alcaraz-Gonzalez V., Mendez-Acosta H., Zitomer D., Totzke D., Spanjers H., Jacobi F., Guwy A., Dinsdale R., Premier G., Mazhegrane S., Ruiz-Filippi G., Seco A., Ribeiro T., Pauss A. and Steyer J. P. (2015). Instrumentation and control of anaerobic digestion processes: A review and some research challenges. *Reviews Environmental Science Biotechnology*, **14**, 615–648, https://doi.org/10.1007/s11157-015-9382-6

Johnson B. R. and Shang Y. (2006). Applications and limitations of ADM 1 in municipal wastewater solids treatment. *Water Science & Technology*, **54**, 77–82, https://doi.org/10.2166/wst.2006.528

Kalyuzhnyi S. and Fedorovich V. (1997). Integrated mathematical model of UASB reactor for competition between sulphate reduction and methanogenesis. *Water Science & Technology*, **36**, 201–208, https://doi.org/10.1016/S0273-1223(97)00524-6

Kalyuzhnyi S. V. and Fedorovich V. V. (1998). Mathematical modelling of competition between sulphate reduction and methanogenesis in anaerobic reactors. *Bioresource Technology*, **65**, 227–242, https://doi.org/10.1016/S0960-8524(98)00019-4

Kalyuzhnyi S., Fedorovich V., Lens P. and Pol L. H. (1998). Mathematical modelling as a tool to study population dynamics between sulfate reducing and methanogenic bacteria. *Biodegradation*, **9**, 187–199.

Kauder J., Boes N., Pasel C. and Herbell J. D. (2007). Combining models ADM1 and ASM2d in a sequencing batch reactor simulation. *Chemical Engineering Technology*, **30**, 1100–1112, https://doi.org/10.1002/ceat.200700045

Kleerebezem R. and van Loosdrecht M. C. M. (2006). Critical analysis of some concepts proposed in ADM1. *Water Science & Technology*, **54**, 51, https://doi.org/10.2166/wst.2006.525

Knobel A. N. and Lewis A. E. (2002). A mathematical model of a high sulphate wastewater anaerobic treatment system. *Water Research*, **36**, 257–265, https://doi.org/10.1016/S0043-1354(01)00209-3

Koster W. and Van Leeuwen H. P. (2004). Physicochemical kinetics and transport at the biointerface: Setting the stage. In: *Iupac Series on Analytical and Physical Chemistry of Environmental Systems*, **9**, 1–14.

Kong X., Wei Y., Xu S., Liu J., Li H., Liu Y. and Yu S. (2016). Inhibiting excessive acidification using zero-valent iron in anaerobic digestion of food waste at high organic load rates. *Bioresource Technology*, **211**, 65–71, https://doi.org/10.1016/j.biortech.2016.03.078

Konhauser K. (2007). Introduction to geomicrobiology. *European Journal of Soil Science*, **58** (5), 1215–1216, https://doi.org/10.1111/j.1365-2389.2007.00943_4.x

Koutsoukos P., Amjad Z., Tomson M. B. and Nancollas G. H. (1980). Crystallization of calcium phosphates. A constant composition study. *Journal of the American Chemical Society*, **102**(5), 1553–1557.

Lawrence A. W. and McCarty P. L. (1965). The Role of Sulfide in Preventing Heavy Metal Toxicity in Anaerobic Treatment. *Journal Water Pollution Control Federation*, **37**, 392–406.

Lauwers J., Appels L., Thompson I. P., Degrève J., Impe J. F. and Van Dewil R. (2013). Mathematical modelling of anaerobic digestion of biomass and waste: Power and limitations. *Progress in Energy and Combustion Science*, **39**, 383–402, https://doi.org/10.1016/j.pecs.2013.03.003

Lin C.-Y. (1992). Effect of heavy metals on volatile fatty acid degradation in anaerobic digestion. *Water Research*, **26**, 177–183, https://doi.org/10.1016/0043-1354(92)90217-r

Lizarralde I., Fernández-Arévalo T., Brouckaert C., Vanrolleghem P., Ikumi D. S., Ekama G. A., Ayesa E. and Grau P. (2015). A new general methodology for incorporating physico-chemical transformations into multi-phase wastewater treatment process models. *Water Research*, **74**, 239–256, https://doi.org/10.1016/j.watres.2015.01.031

Loaec M., Olier R. and Guezennec J. (1997). uptake of lead, cadmium and zinc by bacterial exopolysaccharide. *Water Research*, **31**, 1171–1997.

Lopez-Vazquez C. M., Mithaiwala M., Moussa M. S., van Loosdrecht M. C. M. and Brdjanovic D. (2013). Coupling ASM3 and ADM1 for wastewater treatment process optimisation and biogas production in a developing country: Case-study Surat, India.

Journal Water, Sanitation Hygiene Development, **3**, 12, https://doi.org/10.2166/washdev.2013.017

MacConnell C. B. and Collins H. P. (2009). Utilization of re-processed anaerobically digested fiber from dairy manure as a container media substrate. *Acta Horticulturae*, **819**, 279–286.

Maharaj B. C., Mattei M. R., Frunzo L., van Hullebusch E. D. and Esposito G. (2018). ADM1 based mathematical model of trace element precipitation/dissolution in anaerobic digestion processes. *Bioresource Technology*, **267**, 666–676, https://doi.org/10.1016/j.biortech.2018.06.099

Malik A. (2004). Metal bioremediation through growing cells. *Environment International*, **30**, 261–278, https://doi.org/10.1016/j.envint.2003.08.001

Maurer M. and Boller M. (1999). Modelling of phosphorus precipitation in wastewater treatment plants with enhanced biological phosphorus removal. *Water Science & Technology*, **39**, 147–163, https://doi.org/10.1016/S0273-1223(98)00784-7

Maurer M., Abramovich D., Siegrist H. and Gujer W. (1999). Kinetics of biologically induced phosphorus precipitation in waste-water treatment. *Water Research*, **33**, 484–493, https://doi.org/10.1016/S0043-1354(98)00221-8

Mazumder T. K., Nishio N., Fukuzaki S. and Nagai S. (1987). Production of extracellular vitamin B-12 compounds from methanol by Methanosarcina barkeri. *Applied Microbiology and Biotechnology*, **26**, 511–516, https://doi.org/10.1007/BF00253023

Mbamba C. K., Batstone D. J., Flores-Alsina X. and Tait S. (2015a). A generalised chemical precipitation modelling approach in wastewater treatment applied to calcite. *Water Research*, **68**, 342–353.

Mbamba C. K., Tait S., Flores-Alsina X. and Batstone D. J. (2015b). A systematic study of multiple minerals precipitation modelling in wastewater treatment. *Water Research*, **85**, 359–370.

Morel F. and Morgan J. (1972). Numerical method for computing equilibriums in aqueous chemical systems. *Environmental Science & Technology*, **6**(1), 58–67.

Morse J. W. and Arakaki T. (1993). Adsorption and coprecipitation of divalent metals with mackinawite (FeS). *Geochimica et Cosmochimica Acta*, **57**(15), 3635–3640.

Mu H., Chen Y. and Xiao N. (2011). Effects of metal oxide nanoparticles (TiO2, Al2O3, SiO2and ZnO) on waste activated sludge anaerobic digestion. *Bioresource Technology*, **102**, 10305–10311.

Mudhoo A. and Kumar S. (2013). Effects of heavy metals as stress factors on anaerobic digestion processes and biogas production from biomass. *International Journal Environmental Science Technology*, **10**, 1383–1398, https://doi.org/10.1007/s13762-012-0167-y

Musvoto E. V., Wentzel M. C., Loewenthal R. E. and Ekama G. A. (2000a). Integrated chemical-physical processes modelling – I. Development of a kinetic-based model for mixed weak acid/base systems. *Water Research*, **34**, 1857–1867, https://doi.org/10.1016/S0043-1354(99)00334-6

Musvoto E. V, Wentzel M. C. M. and Ekama G. a. M. (2000b). Integrated Chemical Physical Processes Modelling -II. Simulating Aeration Treatment of Anaerobic Digester Supernatants. *Water Research*, **34**, 1868–1880.

Nopens I., Batstone D. J., Copp J. B., Jeppsson U., Volcke E., Alex J. and Vanrolleghem P. a. (2009). An ASM/ADM model interface for dynamic plant-wide simulation. *Water Research*, **43**, 1913–1923, https://doi.org/10.1016/j.watres.2009.01.012

Ohlinger K. N., Young T. M. and Schroeder E. D. (1998). Predicting struvite formation in digestion. *Water Research*, **32**, 3607–3614, https://doi.org/10.1016/S0043-1354(98)00123-7

Oleszkiewicz J. A. and Sharma V. K. (1990). Stimulation and inhibition of anaerobic processes by heavy metals-A review. *Biology Wastes*, **31**, 45–67, https://doi.org/10.1016/0269-7483(90)90043-R

Oude Elferink S. J. W. H., Visser A., Hulshoff Pol L. W. and Stams A. J. M. (1994). Sulfate reduction in methanogenic bioreactors. *FEMS Microbiology Reviews*, **15**, 119–136, https://doi.org/10.1016/0168-6445(94)90108-2

Papirio S., Frunzo L., Mattei M. R. and Ferraro A. (2017). Sustainable Heavy Metal Remediation, Springer, Berlin, https://doi.org/10.1007/978-3-319-61146-4

Parker W. J. and Wu G.-H. (2006). Modifying ADM1 to include formation and emission of odourants. *Water Science & Technology*, **54**, 111–117, https://doi.org/10.2166/wst.2006.532

Paulo L. M., Stams A. J. M. and Sousa D. Z. (2015). Methanogens, sulphate and heavy metals: A complex system. *Reviwes Environmetal Science Biology*, **14**, 537–553.

Peiris B. R. H., Rathnasiri P. G., Johansen J. E., Kuhn A. and Bakke R. (2006). ADM1 simulations of hydrogen production. *Water Science & Technology*, **53**, 129–137, https://doi.org/10.2166/wst.2006.243

Peng S. and Colosi L. M. (2016). Anaerobic digestion of algae biomass to produce energy during wastewater treatment. *Water Environment Research*, **88**(1), 29–39.

Poinapen J. and Ekama G. A. (2010). Biological sulphate reduction with primary sewage sludge in an upflow anaerobic sludge bed reactor - Part 6: Development of a kinetic model for BSR. *Water Sa*, **36**(3), 203–214.

Preis W. and Gamsjäger H. (2001). Thermodynamic investigation of phase equilibria in metal carbonate–water–carbon dioxide systems. *Monatshefte fuer Chemie/Chemical Monthly*, **132**(11), 1327–1346.

Qiao L. and Ho G. (1996). The effect of clay amendment on speciation of heavy metals in sewage sludge. *Water Science and Technology*, **34**(7-8), 413–420.

Rahaman M. S., Mavinic D. S., Meikleham A. and Ellis N. (2014). Modelling phosphorus removal and recovery from anaerobic digester supernatant through struvite crystallization in a fluidized bed reactor. *Water Research*, **51**, 1–10, https://doi.org/10.1016/j.watres.2013.11.048

Reichert P. (1994). *AQUASIM* – A tool for simulation and data analysis of aquatic systems. *Water Science & Technology*, **30**(2), 21–30.

Ristow N. E., Whittington-Jones K., Corbett C., Rose P. and Hansford G. S. (2002). Modelling of a recycling sludge bed reactor using AQUASIM (vol 27, pg 445, 2001). *Water Sa*, **28**, 111–120.

Rittmann B. E., Banaszak J. E., VanBriesen J. M. and Reed D. T. (2002). Mathematical modelling of precipitation and dissolution reactions in microbiological systems. *Biodegradation*, **13**, 239–250, https://doi.org/10.1023/A:1021225321263

Romero-güiza M. S., Vila J., Mata-alvarez J., Chimenos J. M. and Astals S. (2016). The role of additives on anaerobic digestion: A review. *Renewable and Sustainable Energy Reviews*, **58**, 1486–1499, https://doi.org/10.1016/j.rser.2015.12.094

Rosen C., Vrecko D., Gernaey K. V., Pons M. N. and Jeppsson U. (2006). Implementing ADM1 for plant-wide benchmark simulations in Matlab/Simulink. *Water Science & Technology*, **54**, 11, https://doi.org/10.2166/wst.2006.521

Roussel J. (2012). Metal behaviour in anaerobic sludge digesters supplemented with trace nutrients. PhD Thesis, Univ. Birmingham, United Kingdom.

Roussel J. and Carliell-Marquet C. (2016). Significance of vivianite precipitation on the mobility of iron in anaerobically digested sludge. *Frontiers in Environmental Science*, **4**, 60.

Schwarz A. O. and Rittmann B. E. (2007). A biogeochemical framework for metal detoxification in sulfidic systems. *Biodegradation*, **18**, 675–692, https://doi.org/10.1007/s10532-007-9101-2

Scott W. D., Wrigley T. J., Webb K. M. and Science E. (1991). A computer model of struvite solution chemistry. *Talanta*, **38**, 889–895.

Shakeri S., Gustavsson J., Svensson B. H. and Skyllberg U. (2012). Sulfur K-edge XANES and acid volatile sulfide analyses of changes in chemical speciation of S and Fe during sequential extraction of trace metals in anoxic sludge from biogas reactors. *Talanta*, **89**, 470–477, https://doi.org/10.1016/j.talanta.2011.12.065

Shaojun Z. and Mucci A. (1993). Calcite precipitation in seawater using a constant addition technique: A new overall reaction kinetic expression. *Geochimica et Cosmochimica Acta*, **57**(7), 1409–1417.

Smith S., Takács I., Murthy S., Daigger G. T. and Szabó A. (2008). Phosphate Complexation Model and Its Implications for Chemical Phosphorus Removal. *Water Environment Research*, **80**, 428–438, https://doi.org/10.2175/106143008X268443

Solon K. (2015). IWA Anaerobic Digestion Model No. 1 extended with Phosphorus and Sulfur Literature Review, Division of Industrial Electrical Engineering and Automation Faculty of Engineering, Lund University, 1–16.

Solon K., Flores-Alsina X., Mbamba C. K., Volcke E. I. P., Tait S., Batstone D., Gernaey K. V. and Jeppsson U. (2015). Effects of ionic strength and ion pairing on (plant-wide) modelling of anaerobic digestion. *Water Research*, **70**, 235–245, https://doi.org/10.1016/j.watres.2014.11.035

Stumm W. and Morgan J. J. (1996). Aquatic Chemistry, 3rd ed. John Wiley and Sons, Inc, New York.

Stupperich E. and Kräutler B. (1988). Pseudo vitamin B12 or 5-hydroxybenzimidazolyl-cobamide are the corrinoids found in methanogenic bacteria. *Archives of Microbiology*, **149**, 268–271, https://doi.org/10.1007/BF00422016

Sunda W. G. and Huntsman S. A. (1998). Interactions among Cu2+, Zn2+, and Mn2+ in controlling cellular Mn, Zn, and growth rate in the coastal alga Chlamydomonas. *Limnol. Oceanogr*, **43**, 1055–1064, https://doi.org/10.4319/lo.1998.43.6.1055

Szabó A., Takács I., Murthy S., Daigger G. T., Licskó I. and Smith S. (2008). Significance of Design and Operational Variables in Chemical Phosphorus Removal. *Water Environment Research*, **80**, 407–416, https://doi.org/10.2175/106143008X268498

Tait S., Clarke W. P., Keller J. and Batstone D. J. (2009). Removal of sulfate from high-strength wastewater by crystallisation. *Water Research*, **43**, 762–772, https://doi.org/10.1016/j.watres.2008.11.008

Thanh P., Ketheesan B., Zhou Y. and Stuckey D. C. (2015). Trace metal speciation and bioavailability in anaerobic digestion: A review. *Biotechnology Advances*, **34**, 122–136, https://doi.org/10.1016/j.biotechadv.2015.12.006

Thauer R.K. (1998). Biochemistry of methanogenesis: A tribute to marjory stephenson: 1998 marjory stephenson prize lecture. *Microbiology*, **144**(9), 2377–2406.

Uemura S. (2010). Mineral Requirements for Mesophilic and Thermophilic Anaerobic

Digestion of Organic Solid Waste. *International Journal of Environmental Research*, **4**, 33–40.

Uludag-Demirer S. and Othman M. (2009). Removal of ammonium and phosphate from the supernatant of anaerobically digested waste activated sludge by chemical precipitation. *Bioresource technology*, **100**(13), 3236–3244.

Van Briesen J. M. and Rittmann B. E. (1999). Modeling speciation effects on biodegradation in mixed metal/chelate systems. *Biodegradation*, **10**(5), 315–330.

van Hullebusch E. D., Guibaud G., Simon S., Lenz M., Shakeri Yekta S., Fermoso F. G., Jain R., Duester L., Roussel J., Guillon E., Skyllberg U., Almeida C. M. R., Pechaud Y., Garuti M., Frunzo L., Esposito G., Carliell-Marquet C., Ortner M. and Collins G. (2016). Methodological approaches for fractionation and speciation to estimate trace element bioavailability in engineered anaerobic digestion ecosystems: An overview. *Critical. Reviews Environmental Science and Technology*, **46**, 1324–1366, https://doi.org/10.1080/10643389.2016.1235943

Van Langerak E. P. A., Beekmans M. M. H., Beun J. J., Hamelers H. V. M. and Lettinga G. (1999). Influence of phosphate and iron on the extent of calcium carbonate precipitation during anaerobic digestion. *Journal of Chemical Technology & Biotechnology: International Research in Process, Environmental & Clean Technology*, **74**(11), 1030–1036.

Van Rensburg P., Musvoto E. V., Wentzel M. C. and Ekama G. A. (2003). Modelling multiple mineral precipitation in anaerobic digester liquor. *Water Research*, **37**, 3087–3097, https://doi.org/10.1016/S0043-1354(03)00173-8

Vavilin V. A., Vasiliev V. B., Rytov S. V. and Ponomarev A. V. (1995). Modelling ammonia and hydrogen sulfide inhibition in anaerobic digestion. *Water Research*, **29**, 827–835, https://doi.org/10.1016/0043-1354(94)00200-Q

Wang R., Li Y., Chen W., Zou J. and Chen Y. (2016). Phosphate release involving PAOs activity during anaerobic fermentation of EBPR sludge and the extension of ADM1. *Chemical Engineering Journal*, **287**, 436–447, https://doi.org/10.1016/j.cej.2015.10.110

Warren L. A. and Haack E. A. (2001). Biogeochemical controls on metal behaviour in freshwater environments. *Earth-Science Review*, **54**, 261–320, https://doi.org/10.1016/S0012-8252(01)00032-0

Willet A. I. and Rittmann B. E. (2003). Slow complexation kinetics for ferric iron and EDTA complexes make EDTA non-biodegradable. *Biodegradation*, **14**, 105–121, https://doi.org/10.1023/A:1024074329955

Worms I., Simon D. F., Hassler C. S. and Wilkinson K. J. (2006). Bioavailability of trace metals to aquatic microorganisms: Importance of chemical, biological and physical processes on biouptake. *Biochimie*, **88**, 1721–1731, https://doi.org/10.1016/j.biochi.2006.09.008

Wrigley T. J., Scoit W. D. and Webb K. M. (1992). An improved computer model of struvite. *Talanta*, **39**, 1597–1603.

Xu F., Li Y. and Wang Z. W. (2015). Mathematical modelling of solid-state anaerobic digestion. *Progress in Energy and Combustion Science*, **51**, 49–66, https://doi.org/10.1016/j.pecs.2015.09.001

Ye Z. L., Chen S. H., Wang S. M., Lin L. F., Yan Y. J., Zhang Z. J. and Chen J. S. (2010). Phosphorus recovery from synthetic swine wastewater by chemical precipitation using response surface methodology. *Journal of Hazardous Materials*, **176**(1-3), 1083–1088.

Yekta S. S., Lindmark A., Skyllberg U., Danielsson Å. and Svensson B. H. (2014a). Importance of reduced sulfur for the equilibrium chemistry and kinetics of Fe(II), Co (II) and Ni(II) supplemented to semi-continuous stirred tank biogas reactors fed with stillage. *Journal of Hazardous Materials*, **269**, 83–88, https://doi.org/10.1016/j. jhazmat.2014.01.051

Yekta S. S., Skyllberg U., Danielsson Å., Björn A. and Svensson B. H. (2016). Chemical speciation of sulfur and metals in biogas reactors—Implications for cobalt and nickel bio-uptake processes. *Journal of Hazardous Materials*, **324**, 110–116, https://doi. org/10.1016/j.jhazmat.2015.12.058

Yekta S. S., Svensson B. H., Björn A. and Skyllberg U. (2014b). Thermodynamic modelling of iron and trace metal solubility and speciation under sulfidic and ferruginous conditions in full scale continuous stirred tank biogas reactors. *Applied Geochemistry*, **47**, 61–73, https://doi.org/10.1016/j.apgeochem.2014.05.001

Zaher U. and Chen S. (2006). Interfacing the IWA Anaerobic Digestion Model No.1 (ADM1) with Manure and Solid Waste Characteristics. *Proceeding Water Environmental Federation*, **2006**(9), 3162–3175, https://doi.org/10.2175/193864706783751726

Zaher U., Grau P., Benedetti L., Ayesa E. and Vanrolleghem P. (2007). Transformers for interfacing anaerobic digestion models to pre- and post-treatment processes in a plant-wide modelling context. *Environmental Modelling Software*, **22**, 40–58, https:// doi.org/10.1016/j.envsoft.2005.11.002

Zandvoort B. M. H., Hullebusch E. D. Van, Fermoso F. G. and Lens P. N. L. (2006). Trace Metals in Anaerobic Granular Sludge Reactors: Bioavailability and Dosing Strategies. *Engineering Life Science*, **6**, 293–301, https://doi.org/10.1002/elsc.200620129

Zhang T., Bowers K. E., Harrison J. H. and Chen S. (2010). Releasing phosphorus from calcium for struvite fertilizer production from anaerobically digested dairy effluent. *Water Environment Research*, **82**(1), 34–42.

Zhang W., Zhang L. and Li A. (2015). Enhanced anaerobic digestion of food waste by trace metal elements supplementation and reduced metals dosage by green chelating agent [S, S]-EDDS via improving metals bioavailability. *Water Research*, **84**, 266–277, https:// doi.org/10.1016/j.watres.2015.07.010

Zhang Y., Piccard S. and Zhou W. (2015). Improved ADM1 model for anaerobic digestion process considering physico-chemical reactions. *Bioresource Technology*, **196**, 279–289, https://doi.org/10.1016/j.biortech.2015.07.065

Zhao W. E. and Yang H. (2014). Statistical Methods in Drug Combination Studies, CRC Press, Boca Raton, FL.

APPENDIX

Notation

$K_{A/B,H2S}$ Acid-base kinetic coefficient ($m^3 \, kmol^{-1} \, d^{-1}$)

$K_{H,H2S}$ Henry's law coefficient ($kmol^{-3} \, bar^{-1}$)

$K_{I,H2S}$ Inhibition coefficient by undissociated H_2S ($kmol \, m^{-3}$)

K_{100} Concentration of H_2S or pH at which the uptake rate is decreased 100 times ($kmol \, m^{-3}$)

K_2 Concentration of H_2S or pH at which the uptake rate is decreased 2 times ($kmol \, m^{-3}$)

$K_{a,H2S}$	Acid-base kinetic coefficient (m^3 $kmol^{-1}$)
$k_{dec,i}$	First order decay rate of species i (d^{-1})
$k_l a$	Gas-liquid mass-transfer coefficient (d^{-1})
$k_{m,i}$	Monod maximum specific uptake rate (d^{-1})
k_{p2}	Competitive product inhibition coefficient (kg m^{-3})
$K_{S,j}$	Half saturation value (kg COD_Sm^{-3})
$K_{S_{SO4}}$	Half saturation value for sulfate (kg SO_4 m^{-3})
$P_{gas,H2S}$	Hydrogen sulfide partial pressure (bar)
S_j	Concentration of soluble organic components (kg COD m^{-3})
X_i	Concentration of particulate component i (kg COD m^{-3})
I_{pH}	pH inhibition function
I_{H_2s}	Sulfide inhibition function
$\rho_{A/B}$	Acid-base kinetic rate (kmol m^{-3} d^{-1})
ρ_{gas}	Liquid-gas kinetic gas transfer rate (kmol m^{-3} d^{-1})
$\rho_{decay,i}$	Kinetic rate of bacterial decay (kg COD m^{-3} d^{-1})
$\rho_{growth,i}$	Kinetic rate of microbial growth (kg COD m^{-3} d^{-1})

6

Diverse Types of Biogas Plants and Digestate Post-Treatment Processes

C. Marisa R. Almeida[1], Ishai Dror[2], Mirco Garuti[3], Malgorzata Grabarczyk[4], Emmanuel Guillon[5], Eric D. van Hullebusch[6,7], Andreina Laera[8], Nevenka Mikac[9], Jakub Muñoz[10], Dionisios Panagiotaras[11], Valdas Paulauskas[12], Santiago Rodriguez-Perez[13,14], Stephane Simon[15], Jan Šinko[10], Blaz Stres[16,17,18], Sergej Uštak[10], Cecylia Wardak[4] and Ana P. Mucha[1]

[1]*Interdisciplinary Center of Marine Environmental Research (CIIMAR), University of Porto, Terminal de Cruzeiros do Porto de Leixões, Avenida General Norton de Matos, S/N 4450-208 Matosinhos, Portugal*
[2]*Department of Earth and Planetary Sciences, Weizmann Institute of Science, Rehovot 7610001, Israel*
[3]*Centro Ricerche Produzioni Animali (CRPA), Viale Timavo 43/2, 42121 Reggio Emilia, Italy*
[4]*Department of Analytical Chemistry and Instrumental Analysis, Faculty of Chemistry, Maria Curie Skodowska University, Maria Curie Skodowska 3 Sq., 20-031 Lublin, Poland*
[5]*Université Reims Champagne Ardenne, Institut de Chimie Moléculaire de Reims (ICMR), UMR 7312 CNRS-URCA, Moulin de la Housse, BP 1039, 51687 Reims Cedex 2, France*
[6]*Department of Environmental Engineering and Water Technology, IHE Delft Institute for Water Education, Westvest 7, AX, Delft 2611, The Netherlands*

[7]*Institut de Physique du Globe de Paris, Sorbonne Paris Cité, Université Paris Diderot, UMR 7154, CNRS, F-75005 Paris, France*
[8]*Université Paris-Est, Laboratoire Géomatériaux et Environnement (EA 4508), UPEM, 77454 Marne-la-Vallée, France*
[9]*Division of Marine and Environmental Research, Rudjer Boskovic Institute, Bijenicka 54, Zagreb, Croatia*
[10]*Crop Research Institute, Prague, Czech Republic*
[11]*Laboratory of Chemistry, Department of Mechanical Engineering, Technological Educational Institute (TEI) of Western Greece, M. Alexandrou 1, 263 34 Patras, Greece*
[12]*Institute of Environment and Ecology, Agriculture Academy of Vytautas Magnus University, Studentu 11, LT-53361, Kaunas-Akademija, Lithuania*
[13]*Molecular Biology and Biochemical Engineering Department, Universidad Pablo de Olavide – Ctra. de Utrera, km. 1, Seville, Spain*
[14]*IDENER Research, 24-8 Early Ovington, 41300 La Rinconada, Seville, Spain*
[15]*Université de Limoges, PEIRENE EA 7500 URA IRSTEA, 23 avenue Albert Thomas, 87060 Limoges Cedex, France*
[16]*Biotechnical Faculty, University of Ljubljana, Jamnikarjeva 1010, 1000 Ljubljana, Slovenia*
[17]*Faculty of Geodetic and Civil Engineering, University of Ljubljana, Jamova 2, 1000 Ljubljana, Slovenia*
[18]*Faculty of Medicine, University of Ljubljana, Vrazov trg 2, 1000 Ljubljana, Slovenia*

ABSTRACT

Anaerobic digestion (AD) is a biotechnological process in which organic matter is microbially converted into biogas and digestate. Many parameters affect the underlying microbial processes, including depolymerization of organic compounds, acidogenesis, acetogenesis and methanogenesis, as part of the AD cycle. Optimal concentrations of different nutrients and micronutrients are a prerequisite for optimum microbial growth and metabolism in AD processes. The effluent digestate can be used as a substitute for chemical fertilizers, recycling nutrients to create more sustainable agricultural production systems. Trace elements (TEs) can be transferred to soils during application of digestate as fertilizer, being subjected to environmental influences. To evaluate TEs bioavailability and uptake by plants (which can be transferred to the food chain), TEs leaching processes (which can prevent loss of soils nutrients and run off in ground waters), and TEs effects on soil organisms (which can affect soil fertility and productivity), it is relevant to assess the fate and availability of TEs after land application of digestate. This book chapter provides an overview of different type of biogas plants and digestate post-treatment processes.

Possible physicochemical interactions between digestate and soil components, which influence TEs speciation and availability for biological uptake, are also described. Finally, different TEs fractionation and speciation techniques are extensively discussed to give to the reader a good basis when investigating the fate of TEs in soils after digestate application.

KEYWORDS: anaerobic digestion, digestate, TE analysis, fractionation, speciation, trace element (TE)

6.1 INTRODUCTION

Anaerobic digestion (AD) is an established biotechnological process in which organic matter is broken down by microorganisms in controlled anaerobic conditions (biogas plant – BGP) leading to the production of two valuable products: biogas and digestate.

BGPs can be classified according to the treated organic substrates. The first type is BGP that handles livestock effluents, agricultural residues and agro-industrial by-products (agricultural BGPs). The second type handles organic compounds from sewage sludge and wastes of various origins (waste BGPs), and the third type is referred to landfills that handle solid wastes (Seadi et al., 2008). In most cases, the latter is not used as fertilizer due to the presence of many additional organic pollutants or excessive trace-metal burden. In 2016, 17,662 BGPs and 503 biomethane plants were operating in Europe, showing that the biogas industry is an essential part of European development (EBA, 2017). A general description of the main types of BGPs is summarized in Section 6.2.1.

Digestate is the effluent from BGPs and it is considered a natural fertilizer, mainly used for agricultural land, forest soil, re-cultivation of mining sites and brownfields according to national regulations. Digestate application increases organic matter content in the soil and improves land fertility because it is rich in mineral elements like nitrogen, phosphorus, potassium and other nutrients and micronutrients. Considering the carbon balance, about 65–80% of the carbon fed into an agricultural BGP is converted to biogas while the remaining carbon is largely unavailable for anaerobic microorganisms, consequently it is recovered in digestate and incorporated into the soil via digestate application (Dale *et al.*, 2016), thus increasing the soil organic content.

Following AD, digestate usually undergoes solid–liquid separation. Such post-treatment allows concentrating the coarse solids and organic matter in the solid fraction, whereas most of the mineral nitrogen is kept in the liquid fraction (Tambone *et al.*, 2015) (Section 6.2.2). In fact, both digestate fractions can be used as soil amendment, reducing the use of manufactured/chemical fertilisers and optimizing management costs which promote a circular economy (Delzeit & Kellner, 2013). The high nutrient and organic matter content depends on the composition of the input feedstock and on the efficiency of the biological process (Riva *et al.*, 2016).

Optimal concentrations of different elements, such as Fe, Co, Ni, Zn, Mo, W or Se, hereafter referred as TEs, are a prerequisite for optimum microbial growth and metabolism in AD processes. These elements can already be present in the AD substrate, which will depend on the type of substrate, of digester and of employed digestion procedure (mono or co-digestion) (Ezebuiro & Körner, 2017). On the other hand, TEs could be externally added in digesters, either as single elements or as mixed "cocktails" (Garuti *et al.*, 2018).

During application of digestate as a bio-fertilizer, TEs can be transferred to soils, and, thus, released into the environment. Seasonal changes (temperature, water content, rainfall), as well as soil porosity, carbon content, and mineral fraction, will affect the leaching extent of the deposited digestate. After land deposition, AD digestate is exposed to atmospheric air which affects the redox potential of the digestate matrix as well as the redox state of the TEs. TEs can also be adsorbed on to the charged surface areas of the soil's organic and inorganic particles. Additionally, TEs can be complexed by microbially produced molecules and can be taken up by soil microbial biomass as well as plants. In most cases, these processes take place simultaneously at various rates and in different layers of the soil.

The fate and availability of TEs after digestate land application must be assessed to evaluate TEs bioavailability and uptake by plants (which can be transferred to the food chain), TEs leaching processes (which can prevent loss of soils nutrients and run off in ground waters) and TEs effect for soil organisms (which can affect soil fertility and productivity). This book chapter provides an overview of different type of BGPs and post-treatment processes of digestate. The possible physicochemical interactions between digestate and soils, which influence TEs speciation and their availability for biological uptake, are also described. Finally, different TE fractionation and speciation techniques are discussed.

6.2 DIGESTATE – PRODUCTION AND APPLICATION ON SOIL

Digestate contains more inorganic nitrogen (N), which is more accessible to the plants, than untreated slurry. N-uptake efficiency increases considerably and nutrient losses, by leaching and evaporation, are minimized if digestate is used as fertilizer in line with good agricultural practices.

During AD, organic nitrogen is partially mineralized to ammonium and this transformation is dependent upon feedstock used and upon efficiency of the biological process. For example, a longer hydraulic retention time (HRT) leads to lower organic matter content resulting from more efficient anaerobic digestion and methanogenesis due to decomposition of these organics and the breakdown of the resultant organic nitrogen compounds, too. For instance, in livestock slurries nitrogen is mainly ammonium (e.g., in pig slurry 65–70% of nitrogen is ammonium), whereas in energy crops the total nitrogen is mainly organic.

Subsequently, after AD, the ammonium quota could reach 70–85% in livestock slurries and 30–65% in energy crops (Rossi & Mantovi, 2012).

As mentioned previously, digestate is a product of BGP and inherits its characteristics from the substrates processed during anaerobic digestion. Consequently, depending on the BGP, different digestate, with different composition, are generated, which ultimately can affect TEs fate after soil application. In the next sub-sections a general overview of several BGPs and digestate post-treatment processes is described.

6.2.1 Biogas plants: a general overview

6.2.1.1 Agricultural biogas plants

Agricultural BGPs (Figure 6.1) represented about 69% of the total BGPs in the European Union in 2015 (EBA, 2017). AD substrates from agricultural BGPs derived from livestock effluent (e.g., pigs, cows, horses, poultry slurries and manure), energy crops (e.g., maize, sorghum, triticale, other cereals, grasses, sugar beet), agricultural residues (e.g., straw, stovers) and agro-industrial by-products obtained from manufacturing agro-food sector (e.g., molasses, straw, stovers, olive pomace, tomatoes peel, vegetable and fruits manufacturing residues).

Photo Credit: CRPA

Figure 6.1 Agricultural biogas plant.

An indisputable advantage of agricultural BGPs compared to other BGPs, such as waste BGPs, is that the digestate is usually a high quality organic fertilizer that does not contain excessive amounts of potentially hazardous compounds (Koszel & Lorencowicz, 2015).

6.2.1.2 Waste biogas plants

Waste BGPs (Figure 6.2) are involved in the processing of biological degradable wastes (BDW). BDW is a raw material from various origins and substrate compositions, including the organic fraction of municipal wastes (OFMW), waste from the food industry, retail trade (expired food), certain agricultural waste, sludge from wastewater treatment plants, etc. (Usták *et al.*, 2004). The substrates used in these reactors exhibit non-homogeneous properties. Moreover, they can contain impurities, pathogens, trace metals, some organic pollutants and micro-organic pollutants which create waste management concerns. These concerns limit the potential uses of the digestate as a soil conditioner or fertilizer.

Figure 6.2 Waste biogas plant.

Therefore, it was recommended to incorporate a waste separating line before the AD process (Usták *et al.*, 2005), to homogenize (i.e., softening substrate) and perform sufficient hygiene processes (i.e., inactivation of pathogen by pasteurization) of the raw input materials (Bernstad *et al.*, 2013).

Digestate from waste BGPs can be used as fertilizer or soil conditioner eventually after aerobic post-treatment when element concentrations (Cd, Co, Cr, Cu, Ni, Pb and Zn) in composts, solid and liquid digestates are below permitted thresholds (Knoop *et al.*, 2018; Tampio *et al.*, 2016).

6.2.2 Separation of liquid and solid fraction of digestate

The high water content of the digestate requires transporting high volumes. Separation of digestate into liquid and solid fractions may provide a solution by reducing the volume, which is why it is one of the most common first steps in digestate processing.

The most widely used definition of separation efficiency is the total mass recovery of certain components (total solids, volatile solids, total nitrogen, ammonium, total potassium, total phosphorous, etc.) in the solid fraction as a proportion of the total input (Svarovsky, 1985).

The separation efficiency of these mechanical separators can vary widely, due to differences in the efficiency of the separators and because the separation efficiency is affected by the variable physical and chemical composition of the slurry (i.e., the higher the dry matter content, the higher the proportion of solid phase after separation) (Bauer *et al.*, 2009).

6.2.2.1 Solid digestate

Solid digestate generated from agricultural BGPs represents about 3–15% of the wet weight of unseparated digestate. The total solids content is quite high, generally between 15–30%. Dry matter, volatile solids, carbon and phosphorus – in relation to the mass – are significantly accumulated in the solid phase, the amount depending on feedstock and separation technology used. Potassium, on the other hand, is distributed evenly between the digestate liquid and solid fraction (Rossi & Mantovi, 2012).

6.2.2.2 Liquid digestate

Liquid digestate consist ca. 85–90% of the wet weight of unseparated digestate. Total solids content is generally between 1.5–8% when screw press and belt press are used for separation (Rossi & Mantovi, 2012).

The more soluble forms of nutrients are normally contained in the liquid phase, which is characterized by a high phosphorus, nitrogen (35–90% of the nitrogen in this fraction is NH_4^+) and potassium content. The liquid fraction or its derivatives can function as either inorganic or organic fertilizers and/or soil conditioners, providing renewable substitutes for mineral fertilizers based on fossil resources (Vaneeckhaute *et al.*, 2013).

6.2.3 Technologies for application of digestate on soils

Digestates are used very frequently as fertilizers or soil conditioners, especially those from agricultural BGPs, because they contain plant nutrients such as N, P, K, Ca, Fe, Cu, Mn, Zn, Ni and diverse organic compounds. The stable organic matter in the digestate solid phase is excellent for improving soil properties. Increasing soil aeration can improve harvesting better than intensive fertilization of soils with unsuitable physical properties (i.e., poor aggregation, high compaction, inefficient drainage, shallow ploughing, low humus level, high clay content, etc.). Higher soil porosity ensures sufficient flow of nutrient and water. Before application on soil, the dosage of digestate has to be chosen according to its total nitrogen and phosphorus content, crop nutrient requirements for the expected yield and available amount of nutrients in soils (Lukehurst *et al.*, 2010). The soil properties, mainly linked to the soil composition, pH, Ca, Mg and K content, are another factor that influences the dosage of digestate. Finally, the soil organic matter content and cultivation conditions, mainly referring to the previous crop, soil treatment and irrigation, are complementary factors for assessing the required digestate dosage.

The digestate liquid phase can be applied by means of devices used to spread raw sludge, and the digestate solid phase by means of the same devices used for solid manure application. Some devices are shown in Figure 6.3.

Figure 6.3 Digestate application methods: (A) trailing hose, (B) trailing shoe, (C) injection, and (D) splash plate.

Once applied on the soil surface, the digestate has to be incorporated into the soil during the first 24 h, or alternatively, some special machinery capable of soil injections needs to be used. In fact, for the incorporation of digestate to soil, the best application method is by machinery with soil injectors. If only the solid phase is applied, the digestate incorporation could occur in the course of 48 h.

As discussed above, digestate might contain diverse TEs that, despite being nutrients for plants and soil microorganisms, can be toxic at high concentrations. This will depend on digestate origin and treatment. In fact, origin and treatment of digestate will control the digestate composition and the fate of TEs in it. TEs present in digestate can be transferred to soil during digestate application (Kupper *et al.*, 2014). Trace elements may leach to surface or ground waters and/or be taken up by plants and, consequently, transferred into the food chain. All these mechanisms depend on TE speciation and TE bioavailability. Moreover, once digestate is spread on soils, the fate of TE in the environment is influenced by soil properties (soil physical and chemical characteristics).

6.3 SOIL PHYSICO–CHEMICAL CHARACTERISTICS AND TRACE ELEMENT MOBILITY

Soil is a heterogeneous, complex living ecosystem that represents a unique balance between physical, chemical and biological properties. The colour, moisture content and permeability are major physical characteristics of soils.

In addition to the physical characteristics, a lot of other parameters can be considered for soils characterisation. Electrical conductivity (EC), pH, redox potential (Eh), cationic exchange capacity (CEC), exchangeable potassium (K_{exch}), total nitrogen (N_{tot}), nitrates (NO_3^-), sulfur (S), carbonates ($CaCO_3$, $MgCO_3$), ammonium (NH_4^+), organic carbon (OC), total phosphorous (P_{tot}), available phosphorous (P_{avail}), microbial biomass carbon (MBC), TE, and soil respiration are some of the major physicochemical parameters that are essential for soils characterization (Alef & Nannipieri, 1995; Anderson & Domsch, 1989; Bremner & Mulvaney, 1982; Di Bene *et al.*, 2013; Keeney & Nelson, 1982; McLean, 1982; Moscatelli *et al.*, 2005; Nelson & Sommers, 1982; Olsen & Sommers, 1982; Thomas, 1982).

Minerals are major constituents of soils, mainly originating from weathering of parent material. Quartz, feldspars, carbonates (calcite, dolomite), metal oxides/hydroxides and clay minerals (hydrous aluminosilicates) are some of the most important constituents because they play a vital role in the fertilization of soils through adsorption-desorption processes. However, different kinds of minerals show different adsorption-desorption ability and cation exchange capacity (Churchman & Lowe, 2012).

6.3.1 Total trace elements in soils

TEs occur in soils from the natural environment (the Earth's crust) reaching concentrations lower than 100 mg Kg^{-1} (Hooda, 2010). However, anthropogenic activities contribute to the total TE concentrations in soils. One of the major TE sources contributing to diffuse soil pollution is the agricultural practice of spreading mineral and organic fertilizers and pesticides on soil (Kabata-Pendias & Pendias, 2001; Lukehurst *et al.*, 2010).

Some of TEs can be considered as micronutrients because of their involvement in plant growth and animal nutrition. Such TEs are copper, iron, manganese, molybdenum, sodium, zinc, cobalt, chromium, nickel as well as iodine, selenium, boron, chlorine and fluorine (Bennett, 1993; Johnston, 2005).

6.3.2 Mechanisms regulating trace elements distribution in soils

Many TEs in soil solutions exist as a single oxidation state, however, some of them like selenium, chromium and arsenic can occur in different oxidation states according to the existing redox conditions. Common TE speciation in soil solutions includes hydrated cations, oxyanions, organometallic compounds, inorganic complexes or neutral species (Kabata-Pendias, 2011). In the solid part of the soils, the most immobile fraction of TEs occur in the minerals' structure, whereas the most mobile fraction corresponds to TEs bound on the surface of minerals and organic matter of the soil.

Understanding the mobility of TE in soils and their chemical speciation is important to comprehend whether plants or soil microorganisms can easily uptake TE from soils. Moreover, this information will also allow classification of elements as essential or toxic.

The speciation of TE in soils is strongly affected by the physicochemical properties of soils and the toxicity or fertility of soils is influenced by the interactions between the TEs and the soil components (Kabata-Pendias & Pendias, 2001).

The major processes that determine the distribution of TEs in soils include dissolution-precipitation, ion exchange, adsorption-desorption and absorption-mineralization by the biomass (Figure 6.4).

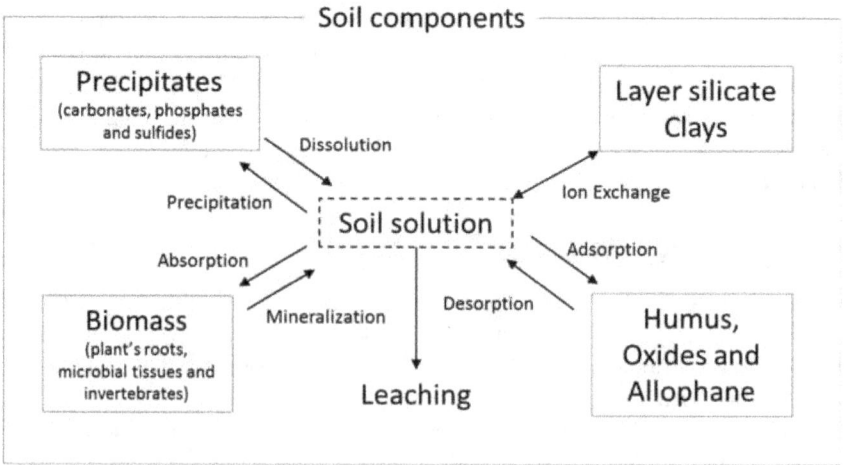

Figure 6.4 Interactive binding mechanisms affecting the distribution of TE between the liquid and solid phases of a common soil (modified from Adriano, 2001).

The distribution of TEs between the solid and liquid phase (i.e., soil solution) depends on soil parameters such as its pH, redox potential (Eh), cation exchange capacity and it is strongly affected by the presence of carbonates, oxides of manganese, iron and aluminium, clays, low-molecular weight organic substances and microorganisms (Kabata-Pendias & Pendias, 2001). The general observed trend is that cations such as cadmium (Cd), copper (Cu), lead (Pb) and zinc (Zn) are dissolved at low pH and Eh, pH having a more important role in TE solubility than Eh (Chuan *et al.*, 1996). Therefore, soils with neutral and alkaline pH behave as a reservoir of TEs that slowly supplies them to plants. On the other hand, in acidic soils TEs are quickly dissolved, thus the TEs are relatively readily available for uptake by plants (Kabata-Pendias & Pendias, 2001).

Another phenomenon which includes the majority of the reactions between TE and the soil components is sorption (Adriano, 2001; Kabata-Pendias & Pendias, 2001), which includes all interaction mechanisms between TE and metals oxides, biomass, carbonates, phosphates, sulphides, basic salts and clays. Moreover, it includes intermolecular interactions (e.g., van der Waals' forces), mechanisms based on valence forces and interface forces such as ion exchange.

A particular sorption process is adsorption, where TEs dissolved in the soil solution interact with the surface of solid particles of soils (Kabata-Pendias & Pendias, 2001). Two types of TE adsorption can be distinguished: a physical adsorption characterized by the weak electrostatic forces where TE are bound through cation exchange, and a surface complexation adsorption, also called chemisorption, which involves stronger chemical bonds between TE and solid particles (Adriano, 2001).

Another binding mechanism is absorption, which involves the assimilation of TE by microorganisms and plants' roots (Adriano, 2001). Indeed, the process of plant transpiration induces the absorption of TE from the soil solution by the roots of plants. In addition, the microbial population surrounding the plant roots will immobilize the TE ions in the microbial biomass. In addition, microbial uptake of TEs is severely affected by positively charged ions supplied along with digestate due to their interference with the direct availability of TEs to microbial and plant siderophores and soil polyaromatic compounds, such as humic acids and lignin residues.

All of the aforementioned binding mechanisms will strongly depend on the composition of the soil and will help to determine the speciation of TEs and their mobility and bioavailability into the environment (Adriano, 2001).

6.3.3 Trace elements mobility in soils and availability for plants

In general, the mobility of selected TE in soils decreases in the following order:

cadmium $\gg\gg\gg$ nickel $\gg\gg$ zinc $\gg$ copper $>$ lead (Tlustoš *et al.*, 2007).

The most important factors regulating TE solubility and availability for plants are the redox potential and pH values. Trace element availability is mostly lower in heavy soils with neutral or alkaline pH values and higher in light soils (Wenzel *et al.*, 1999). Generally, TE content in plant tissues decreases with increasing pH values in plant tissues (Tyler & Olsson, 2002).

Redox potential is mainly important in relation to iron, manganese and elements adsorbed onto Fe and Mn oxy-hydroxides, such as arsenic, copper, mercury and lead. Under reducing conditions, Fe and Mn oxyhydroxides dissolve releasing adsorbed elements into soil solution (Chuan *et al.*, 1996). Elements with higher oxidation numbers are usually less soluble.

Organic compounds participate in the processes that reduce element concentration in soil solutions either by nonspecific or specific sorption (soil ability to bond ions or molecules of different compounds from soil solution into soil solid phases) or by complexation that protect elements against adsorption or precipitation (Alloway, 1990; Shaheen *et al.*, 2013). The solubility of soil organic matter increases with an increase in pH, which influences complex stability consisting of humus and trace elements. Humic acids are insoluble at neutral and acidic pH values and they can cause the retention of harmful TEs in soils, and the decrease of TE concentration in soil solution. Conversely, fulvic acids and their complexes are soluble even at low pH values and they may increase TE mobility, to which they are bonded (Shaheen *et al.*, 2013; Tlustoš *et al.*, 2007).

The cation adsorption of TE depends on the amount of negative electric charge on the soil surface, which is also influenced by pH values. However, some cations can have a higher displacement force than others and they can selectively bond to sorption sites. In addition, competition for sorption sites with more abundant ions, such as calcium and magnesium, will also influence sorption capacity. Cation exchange capacity influences the uptake of TE both directly, by the amount and force of bonded sites for elements in soils, and indirectly, by the influence of other soil properties (Shaheen *et al.*, 2013; Tlustoš *et al.*, 2007).

To summarize, TE mobility and transport in soils and availability to plants will depend on soil physico-chemical characteristic as well as on the TE itself.

6.3.4 Effects of digestates on soil properties and TE content

The digestate application leads to an improvement of soil properties by reducing the bulk density, increasing saturated hydraulic conductivity (water ability to move through pores), raising soil capacity of moisture retention and improving aggregate stability compared with soils without digestate addition. The favourable influence on soil aggregates is caused also because compounds such as lignin, cellulose and hemicellulose complexed with lignin and humic compounds are difficult to degrade during AD, but they are highly reactive with soil surface (Nkoa, 2014). On the other hand, digestates with high sodium content contribute to the dispersion of soil particles.

Soil biota after digestate application is enhanced, especially soil microbial activity (sometimes even macro-organism such as worms are stimulated), because the digestate promotes the activity of some enzymes as urease (decomposes urea to ammonia), phosphatase (important in the phosphorus cycle), β-glucosidonase and dehydrogenase. In addition, the comparisons of parameters describing soil microbial activity, such as basal respiration (the respiration without substrate addition), induced respiration by substrate, specific growth rate, metabolic quotient or nitrogen mineralization capacity, suggest that the effect of digestate application on the promotion of soil microbial activity is lower than the effect of undigested feedstock application in the short-term. Besides, the effects of digestate on different microorganisms can vary.

Digestate compositions combined with soil compositions will affect TE mobility and bioavailability when considering all the mechanisms mentioned in Section 6.3.2 and the organic matter that is being introduced. In fact, once the digestate is applied on soil, TEs will redistribute between the solid and liquid phase of the soil and their mobilization and speciation will change according to the physico-chemical characteristic of the soils and digestates (Guala *et al.*, 2010). Although TE can be already present in the soil, the addition of digestate can lead to the introduction of excess TE. Digestate is produced in anaerobic conditions. However, a fast oxidation of the digestate will occur when it is spread or injected into the topsoil. But, as the digestate moves down through the soil depth, oxic conditions can change to sub-oxic and even anoxic in the deeper soil layers (Borch *et al.*, 2010). This is related to the rate of oxygen penetration and to the soil organic carbon content. Therefore, the solid fraction of digestate can be destabilized and dissolution of TEs can occur together with desorption and/or leaching of metals, a subject that demands further investigation.

So, to fully determine possible TE toxicity during digestate amendment in soil, one must ensure TE analysis not only on the digestate itself but also on soil and on soil amended with digestate, analysing not only total TE content but specially TE speciation and fractionation to establish TE bioavailability. For that one must start with proper sample collection, ensuring sample preservation before analysis.

6.4 COLLECTION, PRESERVATION AND PREPARATION OF SAMPLES

Adequate sampling, sample preservation and preparation are important prerequisites in obtaining representative, reproducible and comparable results of soil analyses (Theocharopoulos *et al.*, 2004) as each of these steps may introduce errors greater than the analytical error.

Collecting, storing and processing samples should keep them as unaltered and free of contamination as possible (Mester & Sturgeon, 2003). van Hullebusch *et al.* (2016) recently reviewed the main methodologies suitable for collecting and preparing anoxic samples. Sampling includes planning of locations, size, number of samples, which should ensure their required quantity and representativeness.

Adequate sampling tools should also be used to obtain undisturbed and contamination-free samples.

To evaluate TE bioavailability and toxicity during digestate application on soil, samples of soil, digestate and soil blended with digestate should be collected and properly preserved and treated. Digestate should be collected before soil application, either liquid or solid fraction depending on what is being applied on soil.

As mentioned previously, digestate, which is in anaerobic conditions in BGPs, is oxidized once outside the reactor and particularly after liquid-solid separation. Therefore, digestate samples can be collected as regular oxic liquid or solid samples according to standard protocols. In contrast, for speciation and fractionation analysis, sampling methods should be carefully evaluated to minimise the conversion of the original chemical species. Furthermore, one should be aware that samples collected should ideally be representative of the whole system. Hence, precautions should be taken to ensure homogeneity of digestate material. With respect to the size of each sample, one needs to consider spatial, that is, the 3D heterogeneity of each soil under analysis. As there are numerous valid approaches to assess the characteristics of soil, one of the most commonly adopted strategies is to distribute the study area into quadrants, sampling a number of cores ($n > 10$) within each of the quadrants and assessing the TE content with depth and location of cores.

The procedure selected for sample preservation depends on the form of the samples stored for analysis (wet or dry) and the parameters to be analysed. Wet samples should be stored deep frozen ($-20°C$ to $-80°C$) and analysed after a short period of time. More often soil samples are dried, either at a room temperature (air dried) in the dark or by freeze drying (lyophilisation) for samples in which volatile compounds can be lost during this step (Mester & Sturgeon, 2003). Dry samples can be stored for several years, at room temperature at a dry place for the total metal analysis or frozen for speciation analysis which includes determination of organometallic compounds, which can be slowly degraded even in dry samples (Dubiella-Jackowska et al., 2007). Sample preparation methods should also be carefully evaluated to minimise the influence on original TE speciation and fractionation.

In general, the soil fraction selected for analysis (total sample of fine fraction) depends on the purpose, but the coarse fraction (>2 mm) is removed from soil samples by sieving after drying the samples up to constant weight. A further step comprises crushing samples (either manually or mechanically) to obtain a homogenous sample with fine particles of uniform size which are suitable for extraction (by organic solvents or leaching reagents) or total dissolution (by mineral acids for total TE analysis) (Mester & Sturgeon, 2003). Although this is valid for total TE content assessment, fractionation and speciation of TE may require that the sample remains at $-20°C$. For instance, the results of fractionation analysis may be affected by drying (humidity level, formation of metal oxides) and grinding (high surface-to-volume ratio, therefore enhanced metal-oxide formation) prior to extraction of different TE fractions (Baeyens *et al.*, 2003). Solid digestate can be collected as a soil sample, with the same concerns being applied.

As for liquid digestate, the total TE content can be determined on the whole sample or for different size fractions, the first involves sample digestion with concentrate acids or simple acid addition, whereas the second can include sequential filtration (van Hullebusch *et al.*, 2016). For TE speciation, different techniques can be applied (see Section 6.5) all including specific sample preparation protocols. In general, the sample should be kept at −20°C.

Regarding soil amendment with digestate, samples can be collected as regular soil samples with the same principles regarding sample collection, preservation and preparation being followed, and taking into consideration further sample analysis. Nevertheless, as mentioned above, as the digestate moves down through soil depth, oxic conditions can change to sub-oxic and even anoxic in the deeper soil layers (Borch *et al.*, 2010). This can change TE mobility and speciation and therefore, soil amended with digestate should be collected over time and at different depths. To ensure the anticipated levels of TE mobility, the oxic/anoxic conditions of the samples need to be maintained comparable to in-situ conditions. In van Hullebusch *et al.* (2016) a review is made on methodologies appropriate for collecting and preparing anoxic samples.

Therefore, standard protocols must be established to set unified procedures for sampling, storing and preparing samples before further analyses, taking in consideration the oxic/anoxic conditions of the samples.

6.5 ANALYSIS OF SAMPLES

In assessing the fate and bioavailability of trace elements in soils after digestate application, one must analyse soil amended with the digestate. Analysis of the digestate (either liquid or solid fraction) and the soil prior to digestate application can also be considered as this will give an indication of the expected total TE concentrations as well as a picture of the changes in TE availability due to digestate application thus allowing for a more proper evaluation of possible TE toxicity and limitations that could be imposed upon digestate application into soil.

As the fate of TEs can be influenced by various parameters such as climate, soil and digestate characteristics as well as the digestate incorporation method (Mantovi *et al.*, 2005) it is crucial to conduct laboratory and field experiments under different but defined conditions. The fate of TEs following short and long-term application of biosolids is relatively well reported in the literature, but this issue remains deeply controversial within the scientific community. Some published data have shown that long-term biosolid spreading would result in the release of metals into the soil due to organic matter mineralization; this is the "time bomb hypothesis" (Bergkvist *et al.*, 2005; McBride, 1995; Stietiya & Wang, 2011). Other authors have suggested that the long-term application of biosolids would present no environmental risk, due to the high adsorption capacity of mineral phases within biosolids; this is the "protection hypothesis" (Hettiarachchi *et al.*, 2006; Li *et al.*, 2001).

Consequently, to better predict the environmental fate and mobility of contaminants, in addition to analysis of the total TE content, determination of TE speciation and fractionation is a prerequisite that ultimately will allow assessment of the possible health risks posed. As indicated by Templeton *et al.* (2000): "TE speciation analysis is the analytical activity of identifying and/or measuring the quantities of individual chemical species" whereas "fractionation analysis refers to separation procedures with insufficient separation power to differentiate between individual chemical species, classifying a group of analytes according to their physical (e.g., size, solubility) or chemical (e.g., bonding, reactivity) properties".

The methodologies commonly used to determine total TEs and TE fractionation and speciation in digestate samples have been recently reviewed by van Hullebusch *et al.* (2016). In the following sub-sections a review of analytical methodologies that can be applied to soil amended with digestate is presented.

6.5.1 Common parameters and total TE concentrations in digestate-amended soils

The general set of sample analyses include measurement of common parameters like pH, EC (electrical conductivity), CEC (cationic exchange capacity), CNS (carbonate, nitrogen and sulphur), organic matter, TOC (total organic carbon), total N, $-NH_4$, total P, total K, temperature, and water content. These analyses are similar to those used for soil prior to digestate amendment and several standard protocols are available.

Total concentration of major (e.g., S, Ca, Mg, Na, Cl, Fe, Mn, Cu, Zn, B) and trace (e.g., As, Co, Cd, Cr, Ni, Pb) elements is often measured and reported (e.g., Alburquerque *et al.*, 2012; Kathijotes *et al.*, 2015; Ramezanian *et al.*, 2015; Szilágyi & Szentmihályi 2009). Atomic absorption (AAS) and atomic fluorescence spectrometry (AFS), as well as inductively coupled plasma spectroscopy (ICP-AES) and spectrometry (ICP-MS), are the main techniques used for total TE concentration determination in soils (Hanlon, 1998; Isaac & Johnson, 1998; Karathanasis & Hajek, 1996).

Electrochemical techniques can also be used for the determination of trace amounts of a wide range of TEs. There are a numerous electrochemical techniques, but they can be subdivided into two groups: potentiometry and voltammetry.

In TE analysis, anodic stripping voltammetry (ASV) has been a popular voltammetric technique because of its speed, good sensitivity and selectivity. In this method, TE is pre-concentrated at the mercury working electrode by the reduction of metal to the metallic form. However, some metals do not form amalgams with mercury. In these cases, adsorptive stripping voltammetry (AdSV) is an adequate technique for their determination in a variety of matrices. In AdSV, high sensitivity can be achieved by adsorptive deposition of a metal complex exploiting the appropriate complexing agent. For instance, for cobalt and nickel dimethylglioxime and nioxime (Ferancova *et al.*, 2016; Korolczuk *et al.*, 2005) can be used, whereas for zinc alizarin is the most suitable choice (Deswati *et al.*, 2016). In the case of selenium, the most useful electrochemical technique is cathodic stripping voltammetry, where high sensitivity is obtained through the formation of an insoluble salt on the mercury electrode followed by its reduction (Grabarczyk & Korolczuk, 2010).

Potentiometry with ion-selective electrodes (ISEs) allows determination of free ion concentrations as well as total metal content in various aqueous samples. The most attractive features of this technique, besides low cost and quick analysis time, are the wide measuring range, the portability of the device, the non-destruction of the sample and the requirement of minimum sample pretreatment. TEs important in AD bioprocesses, for example, cobalt, nickel, zinc and copper, can be determined by ISE (Wardak 2008, 2014; Wardak & Lenik, 2013).

As mentioned above, soil layers with different redox conditions can be encountered and, as such, the oxidation state of the sample should be taken into account upon analysis. Nevertheless, for total TE concentration assessment redox condition is of minor importance.

6.5.2 Fractionation techniques for trace-element partitioning in soils amended with digestate

To investigate the partitioning of TEs in soils, fractionation techniques such as sequential extraction procedures can be adopted. Fractionation can be obtained by wet chemical extraction methods where a series of specific reagents are applied to extract operationally defined species which illustrate elements bound in various soil phases (Tack & Verloo, 1995).

Although the accuracy of sequential extraction methods can be queried due to the numerous extraction steps, operating conditions and reagents involved in extracting TEs, the outcomes of the procedures can provide information on the chemical form of TEs and their mobility in soils (Filgueiras *et al.*, 2002). However, one should be aware that the sequential extraction fractions are operationally defined, that is, the extracted TEs are directly related to the procedure of extraction used (Quevauviller *et al.*, 1997).

In general, sequential extraction procedures consist of consecutive additions of different reagents (extractants) on an aliquot of solid sample to extract diverse fractions of TEs. The extracted fractions have different degrees of mobility into the environment (Filgueiras *et al.*, 2002). Usually, the first extracted fraction corresponds to TEs weakly bound to the solid phase, which have a higher mobility potential compared with later fractions.

Among sequential extraction procedures, the methods most utilized are the ones proposed by Tessier and by the Community Bureau of Reference (BCR). In the Tessier method the fractions are exchangeable, bound to carbonates, bound to iron and manganese oxides, bound to organic matter and the residual fraction (Tessier et al., 1979). Later, the method was modified and adapted for trace-metal extraction in sludge samples in the work of van Hullebusch et al. (2005). This sequential extraction method may be performed in anoxic conditions, which can be important because after digestate amendment different soil layers may have different redox conditions. van der Veen et al. (2007) observed that manganese, nickel and zinc extracted in carbonates and organic matter/sulphide fractions were significantly affected by oxidation, while other metals were not.

Regarding the BCR method, the original method was successively modified (Mossop & Davidson 2003), the fractions being the exchangeable fraction, bound to iron and manganese oxides, bound to organic matter and sulphides and finally the residual fraction (Mossop & Davidson, 2003; van Hullebusch et al., 2005). The performance of the "revised BCR" protocol is, however, more time-consuming compared with the "modified Tessier" protocol.

Despite some drawbacks, such as the uncertainties in the selectivity of the various extractants and the possibility of re-adsorption and partial oxidation of oxygen-sensitive elements (which can be important when dealing with anoxic soil layers), sequential extraction procedures are well established, thus allowing for study of metal partitioning among the various solid phases of soils (Filgueiras et al., 2002).

6.5.3 Speciation of TEs in digestate amended soils

The study of TEs speciation in digestate-amended soils can be achieved by traditional chemical methods such as adsorption/desorption experiments as a function of various parameters for example, pH, contact time, concentration, ionic strength or column experiments in lab or field scale. Such "macroscopic" experiments allow researchers to obtain thermodynamic and kinetic parameters of the sorption processes.

To obtain an in-depth understanding of the processes involved, these experiments can be combined with spectroscopic techniques, such as X-ray absorption spectroscopy (XAS) and X-ray fluorescence microscopy (Donner et al., 2011, 2012). These synchrotron radiation-based techniques are the best available for examination of metal speciation and associations in complex environmental media due to their high resolution and their selectivity for the chosen element. This combination of approaches in term of scale gives useful information to predict metal mobility and bioavailability in biosolid-amended soils. Indeed, the results gained in batch-scale experiments can be compared with those obtained in soil column studies and are then validated with the analytical results attained from the field measurements. If molecular scale studies can also be performed, this enables a profound and general understanding of the fate and behaviour of TEs entering the soil *via* agricultural application of biosolids or being "naturally" present in soils.

As already mentioned above, solid-phase speciation in soil (D'Amore *et al.*, 2005) can be achieved by direct determination of species *in situ* through physical instrumental methods (X-ray methods like XRD, XPD, XAS; magnetic spectroscopies like NMR, EPR; electron techniques like SEM-EDX, TEM-STEM; vibrational spectroscopies like IR, Raman or mass spectrometry like LA-ICPMS). In addition, various hyphenated techniques such as GC-ICP-MS or LC-ICP-MS are used to determine specific compounds of elements like Hg, As, Sb or Cr, which appear in different oxidation states or form stable organometallic compounds in the environment (Bakirdere, 2013; Szpunar, 2000).

Another approach to estimate TE speciation in soil and digestate-amended soil could be through mathematical modelling, using for instance, geochemical equilibrium models. In fact, analytical techniques used to determine TE speciation often are not able to provide information on overall TE speciation. Therefore, mathematical modeling of TE speciation can be used as a theoretical approach to either compile or verify the analytical results.

Whichever methodologies are used, one must be aware that total TE concentrations are not directly linked to TE effects and toxicity and that TE bioavailability is the key factor, a factor that is strongly influenced by TE speciation.

6.5.4 Methods for estimation of TE mobility and bioavailability in soil

6.5.4.1 Diffusive gradient in thin films

In situ determination of bioavailable TE fraction can be estimated by applying passive sampling. The diffusive gradients in thin film (DGT) passive sampling is based on the diffusion of the targeted TE through a diffusive layer (porous gel) followed by an irreversible sorption onto a layer of binding phase. This technique was originally developed to determine labile TE in water (Davison & Zhang, 1994), but its applications have been extended to sediments and soils. During exposure to the device, a concentration gradient of the analyte is established between the exposition media and the binding phase, which results in a flux of analyte correlated with its concentration in the exposition media. After a given time, the accumulated TE is eluted from the binding phase and quantified, usually using AAS or ICP-MS. The original concentration in the exposition media is then determined according to the amount of sorbed TE, the exposition duration and diffusion parameters (TE diffusion coefficient through the diffusive layer, layer thickness) (Davison & Zhang, 1994).

During deployment in a soil, the accumulation of the TEs in the DGT device induces a decrease in the TEs concentration in the soil solution in the vicinity of the device. To compensate this decrease, some TEs can be released from the soil solid phases. As a consequence, the TEs accumulated by the DGT will reflect the initial pore water concentration and its potential resupply from the solid phase (Hooda & Zhang, 2008). Such behavior can be considered as mimicking the uptake of TEs by plants, and thus TE concentration determination by DGT could be more effective in estimating TE bioavailability compared with total analysis or chemical extractions (Sun *et al.*, 2014). Several studies concluded that DGT measurements were actually correlated to the bioavailability of TEs, such as Zn or Cu (Agbenin & Welp, 2012; Almas *et al.*, 2006; Cornu & Denaix, 2006; Quasim *et al.*, 2016; Sun *et al.*, 2014; Tandy *et al.*, 2011; Tian *et al.*, 2008; Wang *et al.*, 2016), but there is no consensus yet. Indeed, poor or no correlation was observed for high TE concentrations (Almas *et al.*, 2006) or during experiments conducted on field cultivation (Agbenin & Welp, 2012). This could be due, for example, to the influence of soil humidity on DGT sampling (Hooda *et al.*, 1999), to competition between major elements of the soil and TEs for sorption onto the binding phase of the DGT device (Mundus *et al.*, 2011), or to the fact that high concentrations of TEs can modify uptake by plants (inhibition of uptake, activation of regulation mechanism, and so on) but not by DGT devices. More studies that include a large variety of TEs, soils and plant species are thus required to evaluate the effectiveness of DGT as a phytoavailability assessment tool.

By taking into account the effective concentration based on pore water and solid-phase releasable amounts, DGT can reveal different rates of release between TEs present in the soil and freshly amended TEs (Zhang *et al.*, 2004). Furthermore, its accumulation capability allows the detection of very low concentrations of labile TE. Finally, DGT can be also used to obtain a depth profile or TE repartition in soils, with an even higher resolution than piezometric sampling (Leermakers *et al.*, 2016; Luo *et al.*, 2013). Such profiles may be helpful to understand the mechanisms of TE release and allow the detection of TE microniches, corresponding to localized highly concentrated mobile TEs, which is not possible with conventional techniques.

6.5.4.2 *Other methodologies*

Biological activities can also influence TE fractionation and speciation through a range of possible mechanisms including precipitation, dissolution, sorption, chelation, and redox transformation. In soils, both plants and microorganisms can have an active role in this. So, TE fractionation and speciation can also be indirectly assessed by specific protocols involving plants and/or microorganisms.

For instance, potential availability of TEs for plant uptake (bioavailable fraction) is usually determined by a single extraction, for which $CaCl_2$, NH_4NO_3 or some complexing agents like EDTA may be used (Chojancka *et al.*, 2005; Pueyo *et al.*, 2004). Other extractants can also be used. For instance, Almeida *et al.* (2005) used low molecular weight organic acids, commonly exuded by plants, to assess metal bioavailability in sediments.

On the other hand, simple toxicity tests can be carried out to evaluate TE effects on soil microorganisms. For instance, commonly used tests are Microtox™ and ToxScreen. ToxSreen is a bioassay that uses a highly sensitive variant of the luminescent bacterium *Photobacterium leiognathi* (Ulitzur *et al.*, 2002). Although these tests are for aqueous samples, they can be also used to test leaching solutions from soils amended with digestate or a slurry of it.

Results obtained through these methodologies should be interpreted jointly with TE fractionation and speciation results to fully assess TE bioavailability and toxicity in sols amended with digestates.

6.6 CONCLUSIONS AND RESEARCH NEEDS

Many countries adopted in their national energy strategies the production of energy from renewable sources, like biogas/biomethane obtained from the anaerobic decomposition of organic substrates, including agricultural residues, biowaste and sludge from sewage treatment.

Digestate is the main out stream from anaerobic digestion process that contains mineral nutrients (nitrogen, phosphorus, potassium) and TEs. The digestate, particularly when it is produced from agricultural substrates, can be used as fertilizer or soil conditioner according to legal requirements. In terms of absorption of nutrients by plants, digestate can resemble mineral fertilizers improving agriculture sustainability.

The analysis of nutrients and TEs in the digestate (e.g., potassium, calcium, iron, copper, manganese, zinc, nickel and chromium), when one considers the digestate as a fertilizer, is required to improve the existing and future digestate application technologies (i.e., precision farming). This knowledge is also necessary to evaluate the cumulative effect of TE content in soil, as well as the soil salinization effect due to excessive content of microelements from the digestate. After application in soil, TE bioavailability will depend on soil characteristics and on digestate composition.

To assess TE bioavailability, TE fractionation and speciation studies should be performed after applying appropriate sampling and sample preservation methods. Several fractionation and speciation methods commonly applied to soil samples can be used for soil amended with digestate. However, the diversity of digestate chemical characteristics and soil chemical properties poses difficulties in setting up an accurate method to investigate the fate of TE in soils amended with digestate.

Long-term field studies should be performed to investigate the fate of TE in soils amended with a large set of digestates. Further research is also needed to assess how much TEs present in digestates may be transferred to plants. It is important to assess how digestate application may affect soil microbial diversity. The outcome of these research works could contribute to improving the knowledge on nutrients and TEs present in anaerobic digestion effluent, and to the establishment of new regulations in the field of organic fertilizers to incorporate digestate, at national and European level.

ACKNOWLEDGEMENTS

This book chapter is based upon work from COST Action 1302 ('European Network on Ecological Roles of Trace Metals in Anaerobic Biotechnologies') supported by COST (European Cooperation in Science and Technology). Czech team's contribution was also supported from the National projects LD15164 and RO0417.

REFERENCES

Adriano D. C. (ed.) (2001). Biogeochemical processes regulating metal behavior. In: Trace Elements in Terrestrial Environments, 2nd edn, Springer, New York, NY, pp. 30–58.

Agbenin J. O. and Welp G. (2012). Bioavailability of copper, cadmium, zinc, and lead in tropical savanna soils assessed by diffusive gradient in thin films (DGT) and ion exchange resin membranes. *Environmental Monitoring and Assessment*, **184**, 2275–2284.

Alburquerque J. A., de la Fuente C., Ferrer-Costa A., Carrasco L., Cegarra J., Abad M. and Bernal M. P. (2012). Assessment of the fertiliser potential of digestates from farm and agroindustrial residues. *Biomass and Bioenergy*, **40**, 181–189.

Alef K. and Nannipieri P. (1995). Methods in Applied Soil Microbiology and Biochemistry. Academic Press, London.

Alloway B. J. (1990). Heavy Metals in Soils. Blackie and Son Ltd., Glasgow and London, 339 pp.

Almas A. R., Lombnæs P., Sogn T. A. and Mulder J. (2006). Speciation of Cd and Zn in contaminated soils assessed by DGT-DIFS, and WHAM/Model VI in relation to uptake by spinach and ryegrass. *Chemosphere*, **62**, 1647–1655.

Almeida C. M. R., Mucha A. P. and Vasconcelos M. T. S. D. (2005). The role of a salt marsh plant on trace metal bioavailability in sediments – estimation by different chemical approaches. *Environmental Science and Pollution Research*, **12**, 271–277.

Anderson T. H. and Domsch K. H. (1989). Ratios of microbial biomass carbon to total organic carbon in arable soils. *Soil Biology and Biochemistry*, **21**, 471–479.

Baeyens W., Monteny F., Leermakers M. and Bouillon S. (2003). Evaluation of sequential extractions on dry and wet sediments. *Analytical and Bioanalytical Chemistry*, **376**, 890–901.

Bakirdere S. (ed.) (2013). Speciation Studies in Soils, Sediments and Environmental Samples, CRC Press, Taylor & Francis, Boca Raton, FL.

Bauer A., Mayr H., Hopfner-Sixt K. and Amon T. (2009). Detailed monitoring of two biogas plants and mechanical solid–liquid separation of fermentation residues. *Journal of Biotechnology*, **142**, 56–63.

Bennett W. F. (ed.) (1993). Nutrient Deficiencies & Toxicities in Crop Plants. APS Press, St. Paul, MN.

Bergkvist P., Berggren D. and Jarvis N. (2005). Cadmium solubility and sorption in a long-term sludge-amended arable soil. *Journal of Environmental Quality*, **34**, 1530–1538.

Bernstad A., Malmquist L., Truedsson C. and la Cour Jansen J. (2013). Need for improvements in physical pretreatment of source-separated household food waste. *Waste Management*, **33**, 746–754.

Borch T., Kretzschmar R., Skappler A., Van Cappellen P., Ginder-Vogel M., Voegelin A. and Campbell K. (2010). Biogeochemical redox processes and their impact on contaminant dynamics. *Environmental Science & Technology*, **44**, 15–23.

Bremner J. M. and Mulvaney C. S. (1982). Nitrogen – total. In: Methods of Soil Analysis, Part 2, Chemical and Microbiological Properties, *Agronomy Monograph*, 2nd edn, Vol. **9**. A. L. Page, R. H. Miller and D. R. Keeney (eds), American Society of Agronomy, Madison, WI, pp. 595–624.

Chojancka K., Chojancki A., Gorecka H. and Gorecki H. (2005). Bioavailability of heavy metals from polluted soils to plants. *Science of the Total Environment*, **337**, 175–182.

Chuan M. C., Shu G. Y. and Liu J. C. (1996). Solubility of heavy metals in a contaminated soil: effects of redox potential and pH. *Water, Air and Soil Pollution*, **90**, 543–556.

Churchman G. J. and Lowe D. J. (2012). Alteration, formation, and occurrence of minerals in soils. In: Handbook of Soil Sciences, 2nd edn, Vol. 1, Properties and Processes, P. M. Huang, Y. Li and M. E. Sumner (eds), CRC Press (Taylor & Francis), Boca Raton, FL, pp. 20.1–20.72.

Cornu J. Y. and Denaix L. (2006). Prediction of zinc and cadmium phytoavailability within a contaminated agricultural site using DGT. *Environmental Chemistry*, **3**, 61–64.

Dale B. E., Sibilla F., Fabbri C., Pezzaglia M., Pecorino B., Veggia E., Baronchelli A., Gattoni P. and Bozzetto S. (2016). Biogasdoneright™: an innovative new system is commercialized in Italy. *Biofuels, Bioprod. Bioref.*, **10**, 341–345.

D'Amore J. J., Al-Abed S. R., Sheckel K. G. and Ryan J. A. (2005). Methods for speciation of metals in soils: a review. *Journal of Environmental Quality*, **34**, 1707–1745.

Davison W. and Zhang H. (1994). In situ speciation measurements of trace components in natural waters using thin-film gels. *Nature*, **367**, 546–548.

Delzeit R. and Kellner K. (2013). The impact of plant size and location on profitability of biogas plants in Germany under consideration of processing digestate. *Biomass and Bioenergy*, **52**, 43–53.

Deswati Pardi H., Suyani H., Zein R., Alif A. and Edelw T. W. (2016). The development method for a sensitive simultaneous determination of Pb(II), Cd(II) and Zn(II) by adsorptive cathodic stripping voltammetry using Alizarin as a complexing agent. *Analytical & Bioanalytical Electrochemistry*, **8**, 885–898.

Di Bene C., Pellegrino E., Debolini M., Nicola Silvestri N. and Bonari E. (2013). Short- and long-term effects of olive mill wastewater land spreading on soil chemical and biological properties. *Soil Biology & Biochemistry*, **56**, 21–30.

Donner E., Howard D. L., De Jonge M. D., Paterson D., Cheah M. H., Naidu R. and Lombi E. (2011). X-ray absorption and micro X-ray fluorescence spectroscopy investigation of copper and zinc speciation in biosolids. *Environmental Science & Technology*, **45**, 7249–7257.

Donner E., Ryan C. G., Howard D. L., Zarcinas B., Scheckel K. G., McGrath S. P., De Jonge M. D., Paterson D., Naidu R. and Lombi E. (2012). A multi-technique investigation of copper and zinc distribution, speciation and potential bioavailability in biosolids. *Environmental Pollution*, **166**, 57–64.

Dubiella-Jackowska A., Wasik A., Przyjazny A. and Namiesnik J. (2007). Preparation of soil and sediment samples for determination of organometallic compounds. *Polish Journal of Environmental Studies*, 16, 159–176.

European Biogas Association (EBA). (2017). The statistical report of the European Biogas Association. EBA, Brussels, Belgium.

Ezebuiro N. C. and Körner I. (2017). Characterisation of anaerobic digestion substrates regarding trace elements and determination of the influence of trace elements on the hydrolysis and acidification phases during the methanisation of a maize silage-based feedstock. *Journal of Environmental Chemical Engineering*, 5, 341–351.

Ferancova A., Hattuniemi M., Sesay A. M., Raty J. P. and Virtanen V. T. (2016). Rapid and direct electrochemical determination of Ni(II) in industrial discharge water. *Journal of Hazardous Materials*, 306, 50–57.

Filgueiras A. V., Lavilla I. and Bendicho C. (2002). Chemical sequential extraction for metal partitioning in environmental solid samples. *Journal of Environmental Monitoring*, 4, 823–857.

Garuti M., Langone M., Fabbri C. and Piccinini S. (2018). Methodological approach for trace elements supplementation in anaerobic digestion: experience from full-scale agricultural biogas plants. *Journal of Environmental Management*, 223, 348–357.

Grabarczyk M. and Korolczuk M. (2010). Development of a simple and fast voltammetric procedure for determination of trace quantity of Se(IV) in natural lake and river water samples. *Journal of Hazardous Materials*, 175, 1007–1013.

Guala S. D., Vegaa F. A. and Covelo E. F. (2010). The dynamics of heavy metals in plant–soil interactions. *Ecological Modelling*, 221, 1148–1152.

Hanlon E. A. (1998). Elemental determination by inductively coupled plasma atomic emission spectrometry. In: Handbook of Reference Methods for Plant Analysis, Y. P. Karla (ed.), CRC Press, Boca Raton, FL, pp. 165–171.

Hettiarachchi G. M., Scheckel K. G., Ryan J. A., Sutton S. R. and Newville M. (2006). μ-XANES and μ-XRF investigations of metal binding mechanisms in biosolids. *Journal of Environmental Quality*, 35, 342–351.

Hooda P. S. (2010). Trace Elements in Soils. John Wiley and Sons, Ltd., London, 618 pp.

Hooda P. S. and Zhang H. (2008). DGT measurements to predict metal bioavailability in soils. *Developments in Soil Science*, 32, 169–185.

Hooda P. S., Zang H., Davison W. and Edwards A. C. (1999). Measuring bioavailable trace metals by diffusive gradients in thin films (DGT): soil moisture effects on its performance in soils. *European Journal of Soil Science*, 50, 285–294.

Isaac R. A. and Johnson W. C. (1998). Elemental determination by atomic absorption spectrometry. In: Handbook of Reference Methods for Plant Analysis, Y. P. Karla (ed.), CRC Press, Boca Raton, FL, pp. 157–164.

Johnston A. E. (2005). Trace Elements in Soil: Status and Management. Essential Trace Elements for Plants, Animals and Humans, NJF Seminar No. 370, 15–17 August 2005, Reykjavík, Iceland, pp. 71–74.

Kabata-Pendias A. (2011). Trace Elements in Soils and Plants. CRC Press, Taylor and Francis, 4th edn. Boca Raton, FL, 534 pp.

Kabata-Pendias A. and Pendias H. (2001). Trace Elements in Soils and Plants, 3rd edn. CRC Press, Boca Raton, FL.

Karathanasis A. D. and Hajek B. F. (1996). Elemental analysis by X-ray fluorescence spectroscopy. In: Methods of Soil Analysis. Part 3. Chemical Methods. Soil Science

Society of America, Book Series Number 5, D. L. Sparks (ed.), American Society of Agronomy, Madison, WI, pp. 161–224.

Kathijotes N., Petrova V., Zlatareva E., Kolchakov V., Marinova S. and Ivanov P. (2015). Impacts of biogas digestate on crop production and the environment: a Bulgarian case study. *American Journal of Environmental Sciences*, **11**, 81–89.

Keeney D. R. and Nelson D. W. (1982). Nitrogen in organic forms. In: Methods of Soil Analysis, Part 2, Chemical and Microbiological Properties. Agronomy Monograph, 2nd edn, Vol. **9**, A. L. Page, R. H. Miller and D. R. Keeney (eds), American Society of Agronomy, Madison, WI, pp. 643–698.

Knoop C., Dornack C. and Raab T. (2018). Effect of drying, composting and subsequent impurity removal by sieving on the properties of digestates from municipal organic waste. *Waste Management*, **72**, 168–177.

Korolczuk M., Moroziewicz A., Grabarczyk M. and Paluszek K. (2005). Determination of Traces of Cobalt in the Presence of Nioxime and Cetyltrimethylammonium bromide by Adsorptive Stripping Voltammetry. *Talanta*, **65**, 1003–1007.

Koszel M. and Lorencowicz E. (2015). Agricultural use of biogas digestate as a replacement fertilizers. *Agriculture and Agricultural Science Procedia*, **7**, 119–124.

Kupper T., Bürge D., Bachmann H. J., Güsewell S. and Mayer J. (2014). Heavy metals in source-separated compost and digestate. *Waste Management*, **34**, 867–874.

Leermakers M., Phrommavanh V., Drozdzak J., Gao Y., Nos J. and Descostes M. (2016). DGT as a useful monitoring tool for radionuclides and trace metals in environments impacted by uranium mining: case study of the Sagnes wetland in France. *Chemosphere*, **155**, 142–151.

Li Z., Ryan J. A., Chen J. L. and Al-Abed S. R. (2001). Adsorption of cadmium on biosolids-amended soils. *Journal of Environmental Quality*, **30**, 903–911.

Lukehurst C. T., Frost P. and Al Seadi T. (2010). Utilization of digestate from biogas plants as biofertilizers. *IEA Bioenergy*, Task 37. 21 pages. Available from: https://www.ieabioenergy.com/publications.

Luo J., Zhang H., Davison W., McLaren R. G., Clucas L. M., Ma L. Q. and Wang X. (2013). Localised mobilisation of metals, as measured by diffusive gradients in thin-films, in soil historically treated with sewage sludge. *Chemosphere*, **90**, 464–470.

Mantovi P., Baldoni G. and Toderi G. (2005). Reuse of liquid, dewatered, and composted sewage sludge on agricultural land: effects of long term application on soil and crop. *Water Research*, **39**, 289–296.

McBride M. B. (1995). Toxic metal accumulation from agricultural use of sludge: are USEPA regulations protective? *Journal of Environmental Quality*, **24**, 5–18.

McLean E. O. (1982). Soil pH and lime requirement. In: Methods of Soil Analysis, Part 2, Chemical and Microbiological Properties. *Agronomy Monograph*, 2nd edn, Vol. **9**., A. L. Page, R. H. Miller and D. R. Keeney (eds), American Society of Agronomy, Madison, WI, pp. 199–224.

Mester Z. and Sturgeon R. (eds) (2003). Sample preparation for trace element analysis. In: Comprehensive Analytical Chemistry, D. Barcello (ed.), Vol. **XLI**, Elsevier, the Netherlands.

Moscatelli M. C., Lagomarsino A., De Angelis P. and Grego S. (2005). Seasonality of soil biological properties in a poplar plantation growing under elevated atmospheric CO_2. *Applied Soil Ecology*, **30**, 162–173.

Mossop K. F. and Davidson C. M. (2003). Comparison of original and modified BCR sequential extraction procedures for the fractionation of copper, iron, lead, manganese and zinc in soils and sediments. *Analytica Chimica Acta*, **478**, 111–118.

Mundus S., Tandy S., Cheng H., Lombi E., Husted S., Holm P. E. and Zhang H. (2011). Applicability of Diffusive Gradients in Thin Films for Measuring Mn in Soils and Freshwater Sediments. *Analytical Chemistry*, **83**, 8984–8991.

Nelson D. W. and Sommers L. E. (1982). Total carbon, organic carbon and organic matter. In: Methods of Soil Analysis, Part 2, Chemical and Microbiological Properties. *Agronomy Monograph*, 2nd edn, Vol. **9**, A. L. Page, R. H. Miller and D. R. Keeney (eds), American Society of Agronomy, Madison, WI, pp. 539–579.

Nkoa R. (2014). Agricultural benefits and environmental risk of soil fertilization with anaerobic digestates: a review. *Agronomy for Sustainable Development*, **34**, 473–492.

Olsen S. R. and Sommers L. E. (1982). Phosphorus. In: Methods of Soil Analysis, Part 2, Chemical and Microbiological Properties. *Agronomy Monograph*, 2nd edn, Vol. **9**, A. L. Page, R. H. Miller and D. R. Keeney (eds), American Society of Agronomy, Madison, WI, pp. 403–430.

Pueyo M., Lopez-Sanchez J. F. and Rauret G. (2004). Assessment of $CaCl_2$, $NaNO_3$ and NH_4NO_3 extraction procedures for the study of Cd, Cu, Pb and Zn extractability in contaminated soils. *Analytica Chimica Acta*, **504**, 217–226.

Quasim B., Motelica-Heino M., Joussein E., Soubrand M. and Gauthier A. (2016). Diffusive gradients in thin films, Rhizon soil moisture samplers, and indicator plants to predict the bioavailabilities of potentially toxic elements in contaminated technosols. *Environmental Science and Pollution Research*, **23**, 8367–8378.

Quevauviller P., Rauret G., López-Sánchez J. F., Rubio R., Ure A. and Muntau H. (1997). Certification of trace metal extractable contents in a sediment reference material (CRM 601). following a three-step sequential extraction procedure. *Science of the Total Environment*, **205**, 223–234. doi:10.1016/s0048-9697(97)00205-2.

Ramezanian A., Dahlin A. S., Campbell C. D., Hillier S. and Öborn I. (2015). Assessing biogas digestate, pot ale, wood ash and rockdust as soil amendments: effects on soil chemistry and microbial community composition. *Acta Agriculturae Scandinavica, Section B – Soil & Plant Science*, **65**, 383–399.

Riva C., Orzi V., Carozzi M., Acutis M., Boccasile G., Lonati S., Tambone F., D'Imporzano G. and Adani F. (2016). Short-term experiments in using digestate products as substitutes for mineral (N) fertilizer: agronomic performance, odours, and ammonia emission impacts. *Science of the Total Environment*, **547**, 206–214.

Rossi L. and Mantovi P. (2012). Digestato, un utile sottoprodotto per il biogas. Centro Ricerche Produzioni Animali, Reggio Emilia, Italy. (In Italian).

Seadi T. A., Rutz D., Prassl H., Köttner M., Finsterwalder T., Volk S. and Janssen R. (2008). Biogas Handbook. University of Southern Denmark Esbjerg, Esbjerg, Denmark.

Shaheen S. M., Tsadilas C. D. and Rinklebe J. (2013). A review of the distribution coefficients of trace elements in soils: influence of sorption system, element characteristics, and soil colloidal properties. *Advances in Colloid and Interface Science*, **201–202**, 43–56.

Stietiya M. H. and Wang J. J. (2011). Effect of organic matter oxidation on the fractionation of copper, zinc, lead, and arsenic in sewage sludge and amended soils. *Journal of Environmental Quality*, **40**, 1162–1171.

Sun Q., Chen J., Ding S., Yao Y. and Chen Y. (2014). Comparison of diffusive gradients in thin film technique with traditional methods for evaluation of zinc bioavailability in soils. *Environmental Monitoring and Assessment*, **186**, 6553–6564.

Svarovsky L. (1985). Solid-Liquid Separation Process and Technology. Handbook of Power

Technology 5. Elsevier, the Netherlands.

Szilágyi M. and Szentmihályi K. (2009). Trace elements in the food chain. Vol. 3. Working Commitee on Trace Elements of the Complex Committe, Hungarian Academy of Sciences (HAS) and Institute of Materials and Environmental Chemistry of the HAS, Budapest, Hungary.

Szpunar J. (2000). Bio-inorganic speciation analysis by hyphenated techniques. *Analyst*, **125**, 963–988.

Tack F. M. and Verloo M. G. (1995). Chemical speciation and fractionation in soil and sediment heavy metal analysis: a review. *Journal of Environmental Analytical Chemistry*, **59**, 225–238.

Tambone F., Terruzzi L., Scaglia B. and Adani F. (2015). Composting of the solid fraction of digestate derived from pig slurry: biological processes and compost properties. *Waste Management*, **35**, 55–61.

Tampio E., Salo T. and Rintala J. (2016). Agronomic characteristics of five different urban waste digestates. *Journal of Environmental Management*, **169**, 293–302.

Tandy S., Mundus S., Yngvesson J., de Bang T. C., Lombi E., Schjoerring J. K. and Husted S. (2011). The use of DGT for prediction of plant available copper, zinc and phosphorus in agricultural soils. *Plant Soil*, **346**, 167–180.

Templeton D. M., Ariese F., Cornelis R., Danielsson L.-G., Muntau H., van Leeuwen H. and Lobinski R. (2000). Guidelines for terms related to chemical speciation and fractionation of elements. Definitions, structural aspects, and methodological approaches. *Pure and Applied Chemistry*, **72**, 1453–1470.

Tessier A., Campbell P. G. C. and Bisson M. (1979). Sequential Extraction Procedure for the Speciation of Particulate Trace Metals. *Analytical Chemistry*, **51**, 844–851.

Theocharopoulos S. P., Mitsios I. K. and Arvanitoyannis I. (2004). Traceability of environmental soil measurements. *Trends in Analytical Chemistry*, **23**, 237–251.

Thomas G. W. (1982). Exchangeable cations. In: Methods of Soil Analysis, Part 2, Chemical and Microbiological Properties. *Agronomy Monograph*, 2nd edn, A. L. Page, R. H. Miller and D. R. Keeney (eds), American Society of Agronomy, Madison, WI, pp. 159–165.

Tian Y., Wang X., Luo J., Yu H. and Zhang H. (2008). Evaluation of holistic approaches to predicting the concentrations of metals in field-cultivated rice. *Environmental Science and Technology*, **42**, 7649–7654.

Tlustoš P., Száková J., Šichorová K., Pavlíková D. and Balík J. (2007). Rizika kovů v půdě v argoekosystémech ČR [Risks of Metals in Soil in Argoecosystems of the Czech Republic]. Vědecký výbor fytosanitární a životního prostředí, Praha. (In Czech).

Tyler G. and Olsson T. (2002). Condition related to solubility of rare and minor elements in fores soils. *Journal of Plant Nutrition and Soil Science*, **165**, 594–601.

Ulitzur S., Lahav N. and Ulitzur N. (2002). A novel and sensitive test for rapid determination of water toxicity. *Environ Toxicol.*, **17**, 291–296.

Uštak S., Váňa J., Kára J., Slejška A., Juchelková D., Žídek M., Šafařík M., Kramoliš P., Muňoz J. and Váňa V. (2004). Anaerobic Digestion of Biomass and Municipal Waste. CZ BIOM & VÚRV, Prague, 116 p. (In Czech).

Uštak S., Váňa J., Kára J., Slejška A., Šafařík M., Kramoliš P. and Vach M. (2005). The Biogas Fermentation of Biomass and Biodegradable Waste. CZ BIOM & VÚRV, Prague, 180 p. (In Czech).

van der Veen A., Fermoso F. G. and Lens P. N. L. (2007). Bonding form analysis of metals and sulfur fractionation in methanol-grown anaerobic granular sludge. *Engineering in*

Life Sciences, **7**, 480–489.

Vaneeckhaute C., Meers E., Michels E., Buysse J. and Tack F. M. G. (2013). Ecological and economic benefits of the application of bio-based mineral fertilizers in modern agriculture. *Biomass and Bioenergy*, **49**, 239–248.

van Hullebusch E. D., Utomo S., Zandvoort M. H. and Lens P. N. L. (2005). Comparison of three sequential extraction procedures to describe metal fractionation in anaerobic granular sludges. *Talanta*, **65**, 549–558.

van Hullebusch E. D., Guibaud G., Simon S., Lenz M., Yekta S. S., Fermoso F. G., Jain R., Duester L., Roussel J., Guillon E., Skyllberg U., Almeida C. M. R., Pechaud Y., Garuti M., Frunzo L., Esposito G., Carliell-Marquet C., Marku C. and Collins G. (2016). Methodological approaches for fractionation and speciation to estimate trace element bioavailability in engineered anaerobic digestion ecosystems: an overview. *Critical Reviews in Environmental Science and Technology*, **46**, 1324–1366.

Wang P., Wang T., Yao Y., Wang C., Liu C. and Yuan Y. (2016). A diffusive gradient-in-thin-film technique for evaluation of the bioavailability of Cd in soil contaminated with Cd and Pb. *International Journal of Environmental Research and Public Health*, **13**, 556.

Wardak C. (2008). Cobalt(II) ion-selective electrode with solid contact. *Central European Journal of Chemistry*, **6**, 607–612.

Wardak C. (2014). Solid contact Zn^{2+}-selective electrode with low detection limit and stable and reversible potential. *Central European Journal of Chemistry*, **12**, 354–364.

Wardak C. and Lenik J. (2013). Application of ionic liquid to the construction of Cu(II) ion-selective electrode with solid contact. *Sensors and Actuators B*, **189**, 52–59.

Wenzel W. W., Lombi E. and Adriano D. C. (1999). Biogeochemical processes in the rhizosphere: In: Heavy Metal Stress in Plants, Role in Phytoremediation of Metal-Polluted Soils, M. N. V. Prasad and J. Hagemeyer (eds), Springer-Verlag, Berlin, pp. 273–303.

Zhang H., Lombi E., Smolders E. and McGrath S. P. (2004). Kinetic of Zn release in soils and prediction of Zn concentration in plants using diffusive gradients in thin films. *Environmental Science and Technology*, **38**, 3608–3613.

7

Potential Recovery Techniques and Digestate Manipulation

*Ana P. Mucha[1], Savic Dragisa[2], Ishai Dror[3],
Mirco Garuti[4], Eric D. van Hullebusch[5,6],
Sabina Kolbl Repinc[7], Jakub Muñoz[8],
Santiago Rodriguez-Perez[9,10], Blaz Stres[7,11,12],
Sergej Uštak[8] and C. Marisa R. Almeida[1]*

[1]*Interdisciplinary Center of Marine Environmental Research (CIIMAR),
University of Porto, Terminal de Cruzeiros do Porto de Leixões, Avenida
General Norton de Matos, S/N 4450-208 Matosinhos, Portugal*
[2]*University of Niš, Faculty of Technology, Bulevar Oslobodjenja 124, 16000
Leskovac, Serbia*
[3]*Department of Earth and Planetary Sciences, Weizmann Institute of
Science, Rehovot 7610001, Israel*
[4]*Centro Ricerche Produzioni Animali (CRPA), Viale Timavo 43/2, 42121
Reggio Emilia, Italy*
[5]*Department of Environmental Engineering and Water Technology,
IHE Delft Institute for Water Education, Westvest 7, AX Delft 2611,
The Netherlands*
[6]*Institut de Physique du Globe de Paris, Sorbonne Paris Cité, Université
Paris Diderot, UMR 7154, CNRS, F-75005 Paris, France*
[7]*Faculty of Civil and Geodetic Engineering, University of Ljubljana, Jamova
2, 1000 Ljubljana, Slovenia*
[8]*Crop Research Institute, Prague, Czech Republic*
[9]*Molecular Biology and Biochemical Engineering Department,
Universidad Pablo de Olavide - Ctra. de Utrera, km. 1, Seville, Spain*
[10]*IDENER Research, 24-8 Early Ovington, 41300 La Rinconada, Seville, Spain*

[11]*Biotechnical Faculty, University of Ljubljana, Jamnikarjeva 101, 1000 Ljubljana, Slovenia*
[12]*Faculty of Medicine, University of Ljubljana, Vrazov trg 2, 1000 Ljubljana, Slovenia*

ABSTRACT

Biogas plants receive inputs of different sources of carbon, nutrients, metals and other pollutants from large areas that result in a digestate that is a very complex and concentrated matrix. How to redistribute all these components without causing imbalances in the receiving environments is one of the main questions that arises regarding the reuse of digestate. The main end destinations of digestate within the EU are agriculture, landfill and incineration, in addition to open-mine land reclamation. There are European and country specific end destinations of digestate that have been recently reviewed and made publicly available in an EU commission report. In terms of agricultural application, digestate is seen as a valuable source of carbon and nutrients, but its application is conditioned by disposal limits for nitrogen, phosphorous and metals. Here, we discuss the need for redesign of the process of digestate manipulation in order to increase its value as fertiliser, through addition of compounds, different solid/liquid phases separation or additional treatments. Potential recovery techniques are also discussed. Phytoremediation, the use of plants to uptake metals from different substrates, can be used not only to remove trace metals from the digestate but also for the recovery of metals from plant biomass or their reintroduction into the biodigester. In addition, a combination of landfill with phytoremediation can be a good alternative for the recovery of degraded soils, or for the reclamation of polluted soil for landscape recovery. Another option can be the use of digestate to produce biochar to be applied in agriculture, a technique that increases carbon content in soils while decreasing trace metal bioavailability. Finally, we discuss the new opportunities that are arising for the use of digestate, including microalgae biomass production and bioenergy.

KEYWORDS: anaerobic digestion, biochar, digestate, recovery, reuse, trace metal

7.1 INTRODUCTION

Digestate is the effluent of the anaerobic digestion (AD) process after recovery of biogas. It can be used as fertiliser on land due to its excellent fertiliser qualities, based on a rich content of plant macronutrients including nitrogen (N), phosphorus (P), potassium (K), and sulphur (S), various micronutrients and also organic matter. AD leads to the reduction of biodegradable organic matter of original substrates, but does not diminish the content of nitrogen, phosphorous, potassium and other nutrients. The content of total solids decreases during AD, so digestate can contain 50–80% less total solids in comparison with the input substrate (Holm Nielsen et al., 1997). One of the most common technologies for agricultural biogas plant is wet anaerobic digestion: the total solids content of digestate in general varies between 2% and 10% depending on feedstocks, operational conditions (temperature, mixing system and tank geometry) and rheological proprieties of the fluid. During the AD process, part of the organic nitrogen is mineralized into ammonium ($N-NH_4^+$) in a way which is dependent on the feedstock used. Digestate is the result of a microbial process and therefore has characteristics that are specific to each digester tank and is also influenced by post-treatment (solid/liquid separation, stripping, evaporation, drying, composting, biological oxidation steps, others).

7.1.1 The complexity of digestate

Biogas plants receive inputs of different sources of carbon, nutrients, metals and other pollutants from large areas that result in a digestate that is a very complex and concentrated matrix. The term 'feedstock' could be defined to include any substrate that can be anaerobically converted to methane; feedstocks suitable for AD are many and varied and many billions of tonnes are available in Europe. Historically AD has mainly been associated with the treatment of animal (pig, cattle, poultry) manure and sewage sludge from aerobic wastewater treatment plants. However, in the 1970s increased environmental consciousness, accompanied by a demand for new waste-management strategies and renewable energy forms, broadened the field of applications for anaerobic digestion and hence introduced additional industrial and municipal wastes (Steffen et al., 1998) (Figure 7.1). Nevertheless, agriculture accounts for the largest potential feedstocks. Mono-digestion of manure and slurries as substrate for AD gives relatively low biogas yields per unit of wet weight; for this reason, frequently agricultural wastes are co-digested with other energy-rich feedstock to provide higher biogas yields (Braun & Wellinger, 2003). Commonly used co-substrates include energy crops for their high content in cellulose and hemicellulose. The use of ensiled plants cereals (maize, sorghum, triticale) for AD in agricultural digesters is a common practice. The silage can be stored over prolonged periods of time and used for biogas production during the year.

The nutrient composition of input substrates is very important to ensure stable process conditions and efficient organic matter degradation. Mainly, C/N ratio of the feedstock influences the growth of microorganisms. A high C/N ratio carries a risk of nitrogen limitation for the growth of microorganisms, buffer capacity limitation in fermentation medium and volatile fatty acids accumulation which can result in low efficiency of degradation (Igoni et al., 2008), while a low C/N ratio may lead to an increase in ammonia concentration, which may inhibit the microbial communities (Rajagopal *et al.*, 2013). The C/N ratio of the organic substrate should be around 15–30:1 (Igoni *et al.*, 2008; Weiland, 2010) to obtain maximum bacterial growth in biogas reactors. However, stable process can be obtained at both lower and higher C/N ratio (Mata-Alvarez *et al.*, 2014; Moestedt *et al.*, 2015; Yan *et al.*, 2015) showing that other operational factors are also relevant.

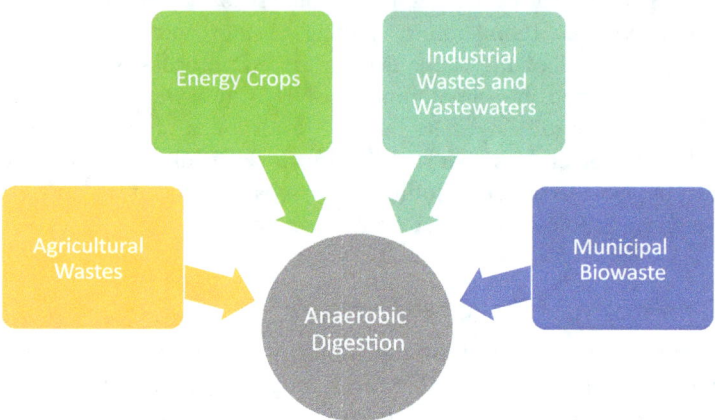

Figure 7.1 Sources of substrates for anaerobic digestion (adapted from Steffen *et al.*, 1998).

The content of carbon in the digestate is generally lower in comparison with the input substrate used as feedstock. This lower carbon concentration in digestate can be explained by the mineralization of carbon in CO_2 and by methane (CH_4) production, both originating from the partial anaerobic degradation of carbon. It is generally accepted that lignocellulosic feedstocks containing relatively large amounts of the structural plant polymer lignin have a very low degradability, while substrate containing fats, sugars and alcohols have very high digestibility. In studies comparing fresh manure with digested manure, carbon losses of up to 25–53% have been reported (Kirchmann & Witter, 1992; Möller & Stinner, 2009).

Digestate application on soils promotes the improving soil structure through input of inert organic matter and fibres (primarily lignocellulose), which contributes to the formation of humus in the medium to long term. Digestate spreading on soil can increase the organic matter content of the soil, which is very important for maintaining soil fertility (Masciandaro & Ceccanti, 1999).

7.1.2 End destination

EU-wide biogas plants receive inputs of different sources of carbon, nutrients, metals and other pollutants from physically large areas, hence the resulting organic mass for transformation to biogas derives from a wide area. As a result of this approach of constant concentration, the digestate is a very complex and concentrated organic-inorganic matrix subjected to microbial analysis. How to redistribute all these components without causing imbalances in the receiving environments is one of the main questions that arises with regard to the reuse of the digestate. The main end destinations of digestate within EU are agriculture (e.g., Spain 30%), landfill and incineration, in addition to open-mine land reclamation.

In fact, digestate could be considered as a resource from which several nutrients could be recycled or recovered, mostly nitrogen, phosphorous and potassium, which are used as fertilisers. The feedstock and the operation process are decisive for the composition and quality of the digestate. So, the feedstock used in the AD process is key to selection of the end destination of the digestate. The separated waste streams such as agriculture biomass or households waste provide clear potential criteria for the end destination. However, industrial waste streams such as sewage sludge or co-digestion of a mixed waste might pose problems for the selection of end destination due to the composition of the digestate. While the richness in nutrients makes the digestate a potential source for the agriculture or horticulture, the presence of heavy metals, pharmaceuticals, nano-particles, pathogens and other micro-organic pollutants might limit this potential end destination. In the European Union, the most extended end destination is agriculture and land application, mainly as a result of the main advantages of closing the nutrients cycle and contribution for carbon sequestration and the associated reduction in atmospheric carbon dioxide levels. According to Dahlin *et al.* (2015), 95% of the digestate produced in the European Union has agriculture as the end destination.

Despite the main end destination for the digestate being fertiliser for agriculture, digestate producers must diversify end destinations according to the digestate properties and operation processes. Other end destinations usually include incineration, landfill and open-mine land reclamation. In addition, other sectors which could be potential end destinations are horticulture, landscaping, cattle raising, fuel materials or alternative building materials.

7.1.2.1 Agriculture

7.1.2.1.1 Valuable source of carbon and nutrients

In terms of agriculture application, digestate is seen as a valuable source of carbon and nutrients, but its application is conditioned by disposal limits for nitrogen, phosphorous and metals. Therefore, it is very important to understand how to get rid of the potentially undesirable or over concentrated compounds so that the valuable part of the digestate can be introduced into agricultural soils. In AD of various biologically degradable substrates, proper handling of digestate is a necessity. Many studies so far have been made in applying digestate to crop fields, determining additional values of fertilising with macronutrients (N, P, K) (Insam *et al.*, 2015), their availability (Teglia *et al.*, 2011) and influence on different soil types by using different frequency of application and composition of digestate to improve the properties of the soil. However, long term applications of biogas digestate on human health and the environment still remain insufficiently explored area (Nkoa, 2014). Most of the biogas plants in the past primarily focused on improving biogas production, neglecting digestate properties. Now, safe and quality digestate is an important factor in fertilisation of crops, otherwise problems with inefficiently anaerobically degraded substrates, inappropriate storage of digestate, problems with odor, toxic compounds, pathogen microorganisms and phytotoxicity may cause negative impacts on soil ecosystems and fertility (Alburquerque *et al.*, 2012).

7.1.2.1.2 Limiting factors for disposal of N, P and trace metals

When combining mineral fertilisers and intensive agriculture, many major threats to soil functions have been recognized (Riding *et al.*, 2015): (i) loss of organic matter; (ii) loss of biodiversity; iii) compaction; (iv) erosion; (v) acidification; and (vi) loss of nutrients through leaching. All this can lead to upsetting or even failing of the ecosystem in arable land. Despite many studies having shown that anaerobic liquid and solid digestate can be as effective as mineral fertilisers (Nkoa, 2014), and in some cases even better than raw manure (Chantigny *et al.*, 2007), there are several environmental risks associated with land application of anaerobic digestate (Nkoa, 2014). These risks include atmospheric pollution (e.g., ammonia and nitrous oxide emission), nutrient pollution and soil contamination, both chemical (phytotoxic compounds and metals) and biological.

Publicly available specification (PAS) BSI PAS 110:2014 specifies upper limits for Cd, Cr, Cu, Pb, Hg, Ni and Zn in digestate, while other trace-element upper safe limits are not determined. Several studies have shown that digestate contains lower levels of heavy metals than laid out in German, British and Spanish standards (Nkoa, 2014), however the long-term application of heavy metals and its accumulation in soil over repeated applications is not known and needs to be investigated in the future. Toxicity, availability for plant uptake and downward mobility are determined by solubility and speciation of trace elements in soils. However, not all trace elements in soils interact with plants, and interaction is further dependent on physical, chemical, microbial and plant factors compounded by stochastic environmental events and cyclic seasonal fluctuations. Higher concentrations of Zn and Se are toxic to plants and animals, excessive dosing or inhomogeneous application to soils may cause soil infertility (Robinson *et al.*, 2009). Contamination of soil by micronutrients can be seen in mobility and higher heavy metal uptake by plant tissues in sandy soils than in clay soils (Liu, 2016), where pH and organic carbon have influence on transfer of heavy metals from digestate to soils. Furthermore, different parts of plants uptake different amounts of heavy metals.

7.1.2.1.3 Effects of long-term use of digestate and other digestate-related products on soil

The versatility and complex composition of digestate and the many different components that interact with the soil upon its application affect a wide range of physical, chemical and biological properties of the soil (e.g., Makádi *et al.*, 2012). Some of the general physical changes include reducing soil bulk density, increasing hydraulic conductivity and moisture-retention stability and aggregate stability (Diacono & Montemurro, 2010; Hargreaves *et al.*, 2008; Möller, 2015). Hereafter are listed the major long-term effects of the digestate on the chemical and physical properties of soil covering: pH, sodicity, nitrogen, macroelements (P, K and Ca), organic matter, trace elements and microbial activity.

Soil pH: Digestates have an alkaline nature with typical pH values of 7.5–9 (e.g., Gómez *et al.*, 2007; Kataki *et al.*, 2017, Möller & Müller, 2012; Pognani *et al.*, 2009), thus an increase in the soil pH should be expected for natural and acidic soils. However, digestate often includes various acidic compounds (Makádi *et al.*, 2012). Polycondensation, connection to organic and inorganic colloids and transformation of these acids can also have an effect on soil chemical properties through impacting soil colloid content that can decrease of soil pH (Tombácz *et al.*, 1998, 1999).

Soil sodicity: In a recent publication, Pawlett and Tibbett (2015) observed a significant increase in soil sodicity (manifested as increase in both available Na^+ and sodium adsorption ratio (SAR)), with an increased digestate application rate in two field experiments on grassland sites in UK. The increased salinity was attributed to the presence of high sodium concentration in food residue which in turn may jeopardize soil structural stability and plant growth if soil continuously receives digestate application. Reported sodium concentrations in digestates are variable and often range between $\sim$500 mg/Kg^{-1} (Alburquerque *et al.*, 2012) and 3100 mg/Kg^{-1} (Vaneeckhaute *et al.*, 2013). In a different study, Kataki *et al.* (2017) used electrical conductivity (EC) values and their increase compared with a control to demonstrate higher salinity that originates from a continuous application of digestate. In this case too, the authors note that the source of digestate has a strong impact on the applied concentration of salts and consequently on the long-term salinization process.

Soil nitrogen: Generally, the digestate application does not cause any significant changes in the total nitrogen. Many publications (e.g., Alburquerque *et al.* 2012) reported that most nitrogen in digestates occurs as inorganic forms, representing mostly $NH_4^+ - N$. This form of N can be easily lost by ammonia volatilization during storage and land spreading due to the alkaline pH of the digestates (Sommer & Husted, 1995). In addition, NH_4^+-N may be nitrified rapidly in soil, this form being highly available to crops but also subjected to leaching through the soil profile, which may result in groundwater pollution. Therefore, storage and land-spreading operations with digestates must be carefully controlled to avoid negative environmental impacts.

Other macroelements (P, K and Ca): Digestate has higher phosphorus (P) and potassium (K) concentration than that of composts (Tambone *et al.*, 2009), therefore it is more suitable for supplementing these missing macronutrients in soils. However, while no significant change in soil available P content is often reported (e.g., Makádi *et al.*, 2012; Möller & Müller, 2012), the K content of soil is reported to increase with digestate application. Moreover, Möller & Müller (2012) note that the shift in pH has a strong impact on the solubility of P and micronutrients. Raising the pH moves the chemical equilibrium toward the formation of phosphate and subsequent precipitation as calcium or magnesium phosphates.

Soil organic matter: Generally, the amounts of organic dry matter and carbon content of the digestate are decreased by the decomposition of easily degradable carbon compounds in the digestors and leads to the increase of more recalcitrant molecules like lignin, cutin, humic acids, steroids and complex proteins (Pognani *et al.*, 2009; Stinner *et al.*, 2008; Tambone *et al.*, 2009). It is further noted that the digestate like many other organic amendments to soil contain surplus of alkali cations (e.g. K^+, Ca^{2+}, Mg^{2+}, Na^+, NH_4^+) over anions ($H_2PO_4^-$, SO_4^{2-}, Cl^-) which are compensated by bicarbonate, carbonate and organic acids which in turn lead to decreased soil acidity (Yan *et al.*, 1996).

Trace elements: Trace metal content of the feedstock usually originates from anthropogenic source and is not degraded during AD. The main origins of the heavy metals are animal-feed additives, food-processing industry, flotation sludge, fat residues and domestic sewage. One example of a report on trace metals originating from digestate application on soil is a study by Makádi *et al.* (2012) that found that Cd, Co, Cu, Ni and Sr content of soil solutions did not change following digestate application, while Zn content decreased significantly, and the amount of manganese (Mn) increased by almost 40%.

Four to seven years studies on applying digested sludge to soils showed that the concentrations of trace metals (Cu, Pb, Zn) in the top layer of soil was increasing, which calls for close monitoring of trace-metal concentrations in soils and plants, or for a change in a policy of application to an on-off strategy in order to retain trace metal concentration below set limits. Furthermore, other pollution risks such as groundwater contamination by trace metals must be also considered, especially when dealing with sandy soils (Liu, 2016). For example, biogas residues mean values of four-year application rates of heavy metals such as Cu, Zn, Cd, Ni and Pb in the study of Odlare *et al.* (2008) were 57–110, 0.1–0.2, 3–6, 3–5, and 7–13 g ha^{-1} year^{-1}, respectively.

Soil microbial activity: When applying digestate to crops, an 11% increase in soil substrate-induced respiration was achieved, indicating an increase in microbial mineralization potential of organic matter. Microbial activity is important as it liberates nutrients from complex organic materials and makes them available to plants and other members of microbial community. Investigators also found increase an in dormant microbial biomass. The use of biogas digestate gave the largest crop yield and higher levels of active microorganisms compared with undigested fertilisers and mineral fertilisers. Digestate increased the substrate-induced respiration, nitrogen mineralization, potential ammonia oxidation and increased the number of active microorganisms (Odlare *et al.*, 2008), showing that application of biogas digestate alleviated much of the limiting factors present in agricultural soils due to long-term exploitation.

7.1.2.2 Incineration and co-incineration

Anaerobic degradation and transformation of organic matter thermodynamically affects the process under which fermentation is favorable only to a limited extent. This in turn leaves a large mass of organic matter that is locked and inaccessible for further anaerobic degradation and can only be processed in the presence of oxygen, either via microbiological pathways or incineration. Incineration results in a large reduction in the volume of the waste. Depending on the possibilities of re-using ashes, the decrease in the amount of material to be landfilled will be of variable importance. Even though investment costs are more intensive than the cost of the other sludge treatment options, units of significant size can balance investment costs, making incineration a technically and economically viable treatment process in highly dense population areas. The combination of different waste streams, municipal solid waste and waste sludges, also enables optimization of incinerator operations. For incineration, the economic value is limited to close proximity of biogas plants to incineration plants and hence is not universally feasible, nor publicly acceptable, despite the fact that ashes can be seen as a valuable byproduct for subsequent extraction of various inorganic compounds in downstream processing units.

After pre-drying, sludges can also be incinerated in cement kilns because they have a high calorific value. Pollutants are stabilised in the clinker which is an interesting way of treating polluted sludges. From an economic point of view these methods of treatment are mainly justified for sludges not permitted for use in agriculture or incineration in municipal solid waste incinerators. The economics of incineration depend to a great extent on auxiliary fuel requirements and, therefore, temperature, dry matter, volatile solids and calorific value are all important parameters to ensure autogeneous combustion. Rheological properties are important as far as the feeding system is concerned. The toxicity of emissions (gaseous, liquid, solid) depends on the presence of heavy metals and organic micropollutants at origin and/or when improper operating conditions occur. When the sludge is digested, the dry solid content (DS) will be reduced by approximately 20%, due to transformation of organics into biogas. However, in order to use digestate for incineration the DS is normally raised to 40–50%. To make a storable product for multipurpose use, for instance as fertiliser, soil conditioner, fuel etc., the DS is raised to 90–95% and granulated, which is most often a cost-ineffective strategy. Consequently, it is no surprise that today incineration is considered as the last method used in the treatment of digestates, either alone or in combination with other wastes. In saying that, treatment by incineration has represented up to 15% of the total mass of sludges treated in Europe (EEA reports) for the past two decades. Trace metals can be recovered from the resulting ashes and returned to anaerobic digestion in the form of specially formulated chemical additives.

7.1.2.3 Landfill and other land reclamation techniques

Landfill disposal of digestate is most limited in EU and hence not a viable large-scale strategy for the massive disposal of digestate in the future. This also holds for parks, land restoration and landscaping, and open-mine reclamation approaches. However, it is still very important to know whether the sludge is consistent enough to be landfilled. Waste-water sludge can contain all the pollutants contained in raw (inflow) waste water, and the content of organic material varies depending on the proportion of the industrial waste water, but usually falls to the range of 60–70%. From this, it follows that dry matter and volatile solids are the most important parameters in sludge characterisation involved in all the application/disposal methods. These can be modified through stabilisation and solid-liquid separation processes, which are operations almost always present in a waste-water treatment system. Additionally, rheological properties are essential in relation to sludge-bearing capacity. The amount of volatile solids has an impact on the development of malodours and process evolution, including biogas production. Trace metals can negatively affect the evolution of biological processes and the quality of the leachate. Therefore, in the process of siting landfills it has always been taken into account that, even in case of the most careful setting and proper operation, some degree of subsurface pollution may occur. This is the reason why geologically vulnerable sites are avoided (karstic areas and gravel terraces forming subsurface aquifer layers) when locating landfill sites and is very similar to the agricultural use of digestates. In this particular mode of digestate disposal, the reuse of trace metals is not possible, however one must bear in mind that the ongoing microbial processes coupled to newly created soil-like environments will continue to actively degrade organic matter and produce a stream of trace-metal and nutrient-contaminated waters under a variety of conditions.

7.1.3 Regulations for digestate disposal

In the previous section, the potential end destinations for the digestate have been shown. The use in agriculture as fertiliser or the land application as soil conditioner have important advantages, such as reducing dependence on chemical fertilisers and peat, and closes the cycle of nutrients and carbon. Good management in the end destination of digestate will reduce the climate change impact of the waste. However, some health and environmental concerns over the amount and composition of digestate to the selected end destination have been identified.

Health, safety and environmental protection must be ensured to avoid the risks described. The European Union is responsible for marking the guidelines to member states in this challenge. The definition of what is considered waste and what is considered non-waste is *a priori* a key aspect of this challenge. Each state must adjust the protection measures necessary to face these risks in its own waste-management scenarios. The feedstock used in digestate production, waste collection, weather, soil composition and hydrology are some of the parameters to take into account to optimize the digestate end destination in each state.

7.1.3.1 European

The European Union has developed use criteria for waste that becomes a product. The use of digestate on the land can be summarized in three strategies: digestate is a product which is used, digestate is a waste which can be used or the use of which is restricted, and digestate cannot be used. The European Union has considered the following legislative framework to provide optimal guidelines in the selection of the end destination for the digestate.

Directive 2008/98/EC (CEU, 2008) on waste introduces the basis of waste management, the definition of waste, reuse and recovery. The communication from the EU Commission on future steps in bio-waste management in the European Union in 2010 analysed the stage implementation of Directive 2008/98/EC on waste and Directive 1999/31/EC (CEU, 1999) on the landfill of waste. The main conclusions of this communication were: improvement of the separate bio-waste collection, prevention of bio-waste production, revision of the Urban Waste Water Treatment Directive 91/271/EEC (CEU, 1991a) to protect EU soils, chase zero landfilling and the optimization of energy recovery to achieve the renewable energy target for 2020 under the Renewable Energy Directive (proposal). In 2012 the Guidance on the interpretation of key provisions of Directive 2008/98/EC on waste showed the advantages of separate collection bio-waste to produce a high-quality digestate.

Directive on nitrates (91/676/EEC) (CEU, 1991b) protects ground and surface water from nitrate pollution which could be associated with digestate end destination. Fertilisers Regulation (EC/2003/2003) (EC, 2003) ensures nutrient content, safety, and environmental acceptability. The Animal By-products Regulation (EC/1069/2009) (EC, 2009) set the instructions for the collection, use, and removal of animal by-products. Regulation EC/834/2007 (EC, 2007) on organic production and labelling of organic products evaluates which digestates are allowed in organic farming production. Directive 2000/76/EC (CEU, 2000) on the incineration of waste limits negative impact from the co-incineration of waste. This directive restricts the operational conditions for waste co-incineration. Incineration as an end destination for digestate, even with energy recovery, could be influenced by this directive.

7.1.3.2 State specific

There are different approaches to determine the status of digestate as a waste or product according to the individual member state legislation.

When the digestate can be used as fertiliser, the regulation of this end destination has three approaches. One describes the requirements for waste to become a product according to a waste law or environmental regulation. This is the situation of member states such as Germany, France, Denmark, and Austria, which regulate a quality or standardization criteria. Another approach is based on the evaluation of digestate and end destination taking account of the characteristics of the soil and application rate, among other parameters, according to recognized protocols and standards. This is the case for the United Kingdom where the Environment Agency for England and Wales defines the end destination for each situation. Finally, the use in agriculture requires previous registration as a fertiliser according to fertiliser regulations. The Czech Republic, Finland, Greece, Hungary, Italy, Latvia, Netherlands, Poland, Portugal, Spain, and Slovenia show this approach.

The animal by-products regulation also applies as a guideline to the digestate production and end destination because these are potential feedstocks and influence digestate composition. Therefore, the end destination of the digestate from mixed bio-waste should be regulated. United Kingdom adopts the AD quality protocol which classifies quality feedstocks from separated bio-waste. Germany includes legal requirements in the waste and fertilisers legislation which identify bio-waste available for use on soil in the Ordinance on the Utilization of Bio-wastes on Land used for Agricultural, Silvicultural and Horticultural Purposes. The Netherlands defines one quality criteria to the end destination of digestate from different bio-waste in its fertiliser legislation. Three different bio-waste streams are identified, compost, sewage sludge, and other bio-waste from industrial processes. Spain does not specifically regulates the end destination of digestate, but legislation on sewage sludge, digested source-separated bio-waste and digestate organic matter from mixed municipal waste define the end destination. The digestate from co-digestion of bio-waste can be used in agriculture, but digestate from mixed municipal waste cannot. In Estonia, the end destination is regulated by waste, fertiliser, and water legislation on to the use of sewage sludge in agriculture is heavily regulated. Slovenia presents a Decree on the treatment of biodegradable waste which regulates the mandatory controls on the feedstock in the digestate production. This regulation identifies a list of suitable bio-waste to be taken into account in the selection of end destination. Austria has a Guideline on the use of digestate on agricultural land according to a positive list of feedstocks which are based on waste-separated collection and uses clean organic sources. The Italian regulation introduces a section dedicated to the agronomic use of digestate from agricultural biogas plants depending on the characteristic of the feedstock used (quality standards of digestate are defined in the regulation).

7.2 REDESIGN OF THE DIGESTATE PROCESS TO INCREASE DIGESTATE VALUE AS FERTILISERS

Characteristics of the digestate are strictly related to feedstock properties and are specific to each digester tank or, even, within the same batch of digestate (Lukehurst *et al.*, 2010). During AD, the carbon content in the digestate is significantly reduced since the organic dry matter is transformed into methane and carbon dioxide. Also, a part of organically bound nitrogen is mineralized and the amount of ammonium in the digestate is higher than in the other organic fertilisers (Roschke & Plöchl, 2006).

During the storage of the digestate a certain amount of ammonia is released into the atmosphere. Also, storage can cause a decrease in the total solids, chemical oxygen demand and alkalinity of the digestate (Laureni *et al.*, 2013). An excess of nutrients present in the digestate can cause environmental problems. So, the digestate needs to be processed in order to manage nutrient content. Removal of particulate nitrogen can be performed by solid-liquid separation, while ammonia removal can be achieved through the use of chemical/physical and biological processes (Silvestri *et al.*, 2013).

Quality of the digestate for use as a fertiliser is defined by nutrient content, pH, dry matter and organic dry matter content, homogeneity, purity (free of inorganic impurities such as plastic, stones, glass, etc.), sanitization and safety for living organisms and the environment with respect to its biological content (pathogenic) material and chemical pollutants (organic and inorganic) (Al Seadi & Lukehurst, 2012). The use of digestate must meet a range of legislative requirements both for agricultural best practice and environmental protection. To increase a digestate value as fertiliser, without adverse impact on methane yield, the following different techniques can be applied before and after the digestion process.

7.2.1 Pre-digestion techniques – Feedstock pre-treatment

Due to its ability to degrade many of unwanted compounds and pollutants within the feedstock, a stable AD process has a positive effect on digestate quality for use as fertiliser. If a digestate is used as fertiliser or for other agricultural purposes, a feedstock must not be used in biogas plants if efficient pollutant removal cannot be guaranteed either by pre-treatment or through the AD process (Al Seadi & Lukehurst, 2012). In order to remove, decompose or inactivate unwanted impurities, the feedstock can be pre-treated by mechanical, chemical and/or thermal techniques. The unwanted impurities or contaminants that influence the quality and safety of digestate used as fertiliser are grouped as: physical impurities (indigestible materials), chemical impurities (trace metals and organic pollutants), and pathogens and other unwanted biological matter (animal and plant pathogens, weed seed).

Despite the fact that the most common pathogens and common viruses are killed during mesophilic and thermophilic digestion, a pre-sanitation step (mostly by batch pasteurization) can be applied for some specific feedstock types, prior to being added to the digester and mixed with the rest of the material. Others pre-treatments for enhancing digestibility of the material include maceration, thermal and chemical hydrolysis, ultra-sound treatments etc., and they are usually applied to materials that contain significant portions of lignocelluloses and hemicelluloses.

7.2.2 Post-digestion techniques

After removal from the digester, digestate can be used as fertiliser without any further treatment. Since storage, transport and application of the digestate are expensive due to low dry matter content, digestate processing is a necessary option for volume reduction and quality enhancement. Digestate processing can be partial (solid-liquid separation, volume reduction), or it can be complete, separating the digestate into solid fibres, fertiliser concentrates and pure water (Al Seadi & Lukehurst, 2012). The aim is to produce a standardized solid or liquid biofertiliser with improved quality (higher concentrations of plant nutrients than unprocessed digestate, separate nutrients in mineral form) and marketability.

7.2.2.1 Solid-liquid separation

The first step in digestate processing is to separate the solid phase from the liquid phase. Digestate separation techniques have been divided into categories based on the type of process employed, that is, mechanical, thermal (evaporation) or biological (bio-drying), or a combination of these. Efficiency of separation essentially depends upon the nature of the digestate and the characteristics of particles. Different methods can be used for mechanical liquid separation (Lukehurst *et al.*, 2010; Pöschl *et al.*, 2010) including: belt press, screw press, sieve drum, sieve centrifuge, decanter centrifuge. Bauer *et al.* (2009) indicated that a screw separator is more suitable for separation comparing with a rotary screen separator. In the same research, the dry matter content in the liquid fraction was 4.5% and in the solid fraction 19.3%. After filtering through the pore size under 0.5 mm a significant enrichment of nutrients in the solid phase can be expected (Møller *et al.*, 2000). Depending on the method efficiency, the separation of dry matter, phosphorus, nitrogen and potassium can vary (Lukehurst *et al.*, 2010). The dispersion of the nutrients between the liquid and solid fraction is different, so the liquid fraction has more nitrogen and potassium, while the solid fraction contains volatile solids, carbon, raw ash and phosphorus (Bauer *et al.*, 2009; Liedl *et al.*, 2006). For digestate evaporation, the heat is sourced from the gas engine surplus heat in order to make the process financially sustainable. Bio-drying refers to the removing of water by the composting process, i.e using aerobic bacteria to heat the digestate and remove the water (Al Seadi & Lukehurst, 2012).

Solid-liquid separation has several advantages (Lukehurst *et al.*, 2010; Wu *et al.*, 2016): (i) the volume of the required storage tank is reduced; (ii) digestate is separated into a stackable dry fraction and a pumpable liquid fraction; (iii) nitrogen uptake is more efficient from the liquid fraction; (iv) liquid fraction can be recirculated into the digestor; and (iv) there is little need for mixing of the liquid phase before the spreading.

The drawback of the use of liquid phase is in uneconomical transportation due to the high content of water and low efficiency compared to chemical fertilisers (Möller & Müller, 2012). The solid fraction can be used directly after separation, or can be dried or composted (Pöschl et al., 2010). Since the solid fraction of digestate is considered as a waste in order to be marketed and used it can be composted (Tambone et al., 2015). The aim of the composting is to obtain a stable and mature compost that can be easily stored and handled (Himanen & Hänninen, 2011), but some results (Tambone et al., 2015) showed that the process does not significantly improve the characteristics of the solid fraction of the digestate.

Separation can be improved by the use of chemicals for coagulation or flocculation of liquid before the centrifugation (Lukehurst et al., 2010). After the separation of digestate, complete conditioning of digestate can be performed in order to get three final products: water, concentrated mineral nutrients and organic fibres. Conditioning of digestate can be performed by the use of membrane separation and evaporation (Lukehurst et al., 2010). Other techniques that can be used are microfiltration, ultrafiltration and reverse osmosis (Ledda et al., 2013; Silvestri et al., 2013). As a result of the separation by ultrafiltration and reverse osmosis, a nutrient-concentrated fertiliser rich in organic compounds and decontaminated water can be obtained. If the process is effective, the water quality can be similar to potable water (Silvestri et al., 2013). Despite being the most expensive technology, membrane purification is among the most frequently applied approaches in more complex digestate processing facilities in Germany, Switzerland, and Austria (Drosg et al., 2015).

Changes in pH value can shift the ammonium ion/ammonia equilibrium. Acidification of the digestate can cause the capture of nitrogen in the form of ammonium salts and reduce nitrogen loss after the application of digestate to the land. Increasing the pH value neutralizes odours and reduces the levels of pathogenic microorganisms. Alkaline stabilization is usually achieved by the addition of lime (WRAP, 2012; Silvestri et al., 2013).

7.2.2.2 Digestate recirculation

Digestate recirculation in the AD plant is an interesting possibility which can produce more biogas and reduce unwelcome greenhouse gases emissions (methane and carbon dioxide) into the atmosphere. The residual biomethane potentials of digestate are not only dependent on feedstock, but also upon the hydraulic retention time in the digester. The results of the several studies (reviewed in Monlau et al., 2015) have shown a very high range of values of residual potentials of digestate.

Post-treatment of the entire digestate (mechanical, thermal, thermochemical or enzymatic) or solid digestate prior to recirculation is necessary in order to enhance the methane production of digestate and to improve economic effects. The aim is to enhance the biodegradability of hard-to-digest compounds present in solid digestate.

Two major economic benefits presented by post-treatment of digestate are enhanced process efficiency (an increase of methane yield) and lower cost of post-treating digestates compard with pre-treating feedstock (Monlau *et al.*, 2015). Moreover, reactor perfomance can be improved by enhancing microbial population, since washed-out microorganisms are reintroduced into the process.

7.3 POTENTIAL RECOVERY TECHNIQUES

Biogas digestates have been predominantly used for agricultural soils application. However, the significant increase of digestate production generates problems related to transport costs, greenhouse gas emissions during storage as well as high nitrogen concentrations that restrict its use to land application only (Monlau *et al.*, 2015). Accordingly, different options of biogas digestate reuse are currently under development, such as the use of solid digestate for energy production through biological (e.g., anaerobic digestion) and thermal processes (i.e., combustion, hydrothermal carbonization (HTC) and pyrolysis) (i.e., combining biological and thermochemical processes to obtain higher bioenergy recovery) (Feng & Lin, 2017; Lü *et al.*, 2018; Monlau *et al.*, 2015, 2016) and the conversion of solid digestate into added-value products (e.g., biochar) through thermochemical processes as discussed in the next section.

7.3.1 Biochar digestates

According to the definition given by the International Biochar Initiative, biochar is 'a solid material obtained from thermochemical conversion of biomass in an oxygen-limited environment' (Lehmann & Joseph, 2015). Biochar may be produced from forest or agricultural residues. Biochar production from biogas digestates has emerged as a new valorisation approach over the last decade. Dry feedstock, for example, with moisture content below 50 w%, is generally converted by slow pyrolysis (350–600°C) and wet feedstock by HTC (180–250°C in water above saturated pressure). Biochar properties are affected by the original feedstock physicochemical characteristics and thermal treatment conditions (Hung *et al.*, 2017; Mumme *et al.*, 2011; Stefaniuk & Oleszczuk, 2015). Generally, biochar is rich in stable aromatic carbon and nutrients, making it an eco-friendly material in several ways for soil improvement, mitigation of climate change, nutrient/contaminant pollution, waste management, and energy production (Lehmann & Joseph, 2015; Monlau *et al.*, 2015; Mumme *et al.*, 2011). The use of biochar as an adsorbent for soil and water treatment of organic and metal pollutants has rapidly emerged as a low-cost option (Ahmad *et al.*, 2014; Inyang

& Dickenson, 2015; Laird, 2008; Mohan *et al.*, 2014). More recently another end-use is biochar admixing (as an additive) during anaerobic digestion which has been shown to improve process stability and biogas production (Fagbohungbe *et al.*, 2017; Mumme *et al.*, 2014; Wambugu *et al.*, 2019).

When considering the type of organic waste streams processed, increasing attention has been given recently to finding alternative options in handling biogas digestates, as it is an important aspect for the sustainable development of biogas projects while improving the economic profitability of anaerobic digestion plants. Biochar production from biogas digestate has been reported by several authors (e.g., Hung *et al.*, 2017, Luz *et al.*, 2018; Wongrod *et al.*, 2018a, and many others). For instance, thermochemical conversion of biogas digestate to biochar and its subsequent application to soil as a mechanism to enhance water and nutrient retention is becoming a widely accepted practice (Kataki *et al.*, 2017; Monlau *et al.*, 2016; Mumme *et al.*, 2011; Nansubuga *et al.*, 2015). Monlau *et al.* (2016) reported that solid-digestate and biochar digestate showed good soil amendments properties but with complementary effects. Biochar digestates may act as a source/sorbent of nutrients in soils. Such features have been studied for nitrogen compounds (Takaya *et al.*, 2016; Zheng *et al.*, 2018) as well as phosphorus compounds (Bekiaris *et al.*, 2016; Bruun *et al.*, 2017; Takaya *et al.*, 2016). Biochar digestate may be used for organic and inorganic contaminant immobilization encountered in contaminated soils and water streams. Several authors investigated the performance of biochar digestates in sorbing inorganic pollutants such as Pb and As (Wongrod *et al.*, 2018a, b; Wongrod *et al.*, 2019), Cu and As (Jiang *et al.*, 2018), Cu, Pb and Zn immobilization in industrial soil (Gusiatin *et al.*, 2016) and organic pollutants such as antibiotics (e.g., tetracycline) (Fu *et al.*, 2017; Jiang *et al.*, 2018) and herbicides (e.g., isoproturon) (Eibisch *et al.*, 2015). Biochar digestate application on agricultural land may also contribute to mitigation of climate change. For instance, Schouten *et al.* (2012) reported that biochar produced from cattle manure digestate contributes to decreased CO_2 emission and stabilized N_2O gas emissions when compared with raw cattle manure and anaerobically treated cattle manure digestate when spread onto soils.

Biochar digestates may be used as additives in anaerobic digesters. Anaerobic digestion is performed using microbial consortia that harbour acid producers, which convert substrate to desired acetic acid, CO_2, and H_2, and undesired volatile fatty acids (VFA). The desired products are converted to biogas by the methane producers (Gerardi, 2003). Anaerobic digestion may suffer from low process stability as microbes are sensitive to inhibition. During substrate-induced inhibition, microbes are either inhibited: (1) directly, by toxic substrate fractions (e.g., lipids, metals, pesticides); or (2) indirectly, by toxic degradation products, for example, VFAs, which lower pH as they accumulate, eventually inhibiting the methane producers (Fagbohungbe *et al.*, 2017).

When admixed in anaerobic digestion, biochar can: (1) adsorb direct and indirect toxic compounds; (2) buffer against the increasing VFA since biochar is often alkaline; and (3) provide a surface to immobilize microbes by forming a biofilm (Fagbohungbe *et al.*, 2017). Positive synergistic effects were firstly reported from admixing activated carbon with trace elements (Capson-Tojo *et al.*, 2018, 2019; Zhang *et al.*, 2018). Thus, system integration of biogas and biochar looks promising to take advantage of several profitable synergies (Fagbohungbe *et al.*, 2017; Luz *et al.*, 2018).

7.3.2 Reclamation (landscape recovery)

Reclamation is focused on areas disturbed by human activities, with the intention of returning these areas to optimal conditions of sustainable natural or human-influenced environment. The key priority of reclamation is not oriented to achieve the maximum amount of crop production on the reclaimed areas, but to achieve ecological landscape stability through colonization by microorganisms, plants and animals.

Unique methodology developed for reclamation includes the addition of reclamation compost or similar well stabilized organic or organic-mineral substrate (including aerobically stabilized separated digestate) into the soil production (Ust'ak *et al.*, 2010). This addition, in large quantities, typically from 800 to 1200 t/ha occurs only once at biological recultivation, for reasons of rapid topsoil recovery. Thereafter in the coming years, the soil is treated by conventional methods according to the current applicable regulations in the field of organic fertiliser use (including digestates) during agricultural production (Ust'ak *et al.*, 2010).

Reclamation focused on agricultural activity is based on the cultivation of so-called 'fertilising plants' in a modified crop rotation, leading to the enhancement of soil organic matter provisions and to optimize soil structure. For example, reclamation of areas devastated by mining activities can be achieved by using appropriate agrotechnical methods and crop rotations for agricultural reclamation (Čermák *et al.*, 2002).

7.3.3 Phytoremediation

Clean-up of metal-contaminated soils is truly indispensable due to the metals possible toxic effects. Different physical, chemical and biological methodologies have been employed for this clean up. In general, physical and chemical methodologies have some limitations such as high cost, intensive work, modifications in soil properties, and some of them are irreversible, and can have negative effects on native soil microflora (Ali *et al.*, 2013).

An alternative to physical and chemical methodologies is the use of phytoremediation. This biological methodology has been a promising approach to cleaning up metal-contaminated soils, namely through extraction (phytoextraction), stabilization (phytostabilization), and/or transformation (phytovolatilization) processes (Ma *et al.*, 2016).

Phytoremediation basically refers to the use of plants (and microorganisms associated to plants rhizosphere) to reduce/eliminate contaminants from different environmental compartments (Ali *et al.*, 2013). In fact, the term "phytoremediation" combines two words, the Greek word *phyto* meaning 'plant' and Latin word *remedium* meaning 'to correct or remove an evil'.

This methodology takes advantage of a variety of plant biological processes to support in-site remediation (Pivetz, 2001). It is an innovative, cost-effective, environmental friendly methodology that can be applied in situ, being a solar-driven remediation strategy. It can be applied at very large field sites where other remediation methods are not cost effective or feasible. In general, phytoremediation methodologies have lower installation and maintenance costs than other remediation techniques. In fact, it has been indicated that phytoremediation can cost as little as 5% of alternative remediation methods (Ali *et al.*, 2013, and reference therein). Moreover, vegetated areas are more resistant to erosion and, in the case of metal-contaminated soils, vegetation can also prevent metal leaching. Phytoremediation also has great acceptability for the general public as a "green clean" alternative to chemical facilities and bulldozers (Ali *et al.*, 2013, and references therein).

Among the different phytoremediation techniques, phytoextraction is the most suitable to be used in the removal of metals from contaminated soil and water. Phytoextraction implies the accumulation of metals in harvestable plant biomass that is, aboveground plant shoots (Ali *et al.*, 2013). Phytoextraction includes contaminant uptake by plant roots followed by metal translocation to the aboveground portion of plants and, generally, followed by harvesting and disposal of plant biomass (Pivetz, 2001).

The efficiency of phytoextraction depends on many factors, like bioavailability of the metals in soils. For instance, strong binding of metals to soil particles or metal precipitation can significantly reduce metal availability and therefore significantly reduce metal uptake by plants (Ali *et al.*, 2013, and references therein). However, plants have developed certain mechanisms for increasing metals bioavailability in soil. Plant roots can exude metal-mobilizing compounds in the rhizosphere, for instance phytosiderophores or low molecular weight organic acids (Rocha *et al.*, 2014, 2016). Moreover, microorganisms present in the plant rhizosphere (mainly bacteria and mycorrhizal fungi) may significantly increase the bioavailability of metals in soil (Ma *et al.*, 2016, and references therein). One should be aware that rhizosphere microbial communities also have a key role in phytoremediation. In fact, microorganisms can enhance the phytoremediation potential of a plant in different manners: by promoting plant biomass, increasing (phytoextraction) or decreasing (phytostabilization) metal availability in soil, as well as facilitating metal translocation from soil to root (bioaccumulation) or from root-to-shoot tissues (translocation) (Ma *et al.*, 2016; Oliveira *et al.*, 2014; Rajkumar *et al.*, 2012; Silva *et al.*, 2014).

Plant-root morphology and length are plant characteristics that are also important for phytoremediation. For instance, a fibrous root system with numerous fine roots spread throughout the soil will provide higher contact with the soil due to the higher surface area of the roots. In general, plants have root zones limited to the top layer of soil, which may restrict the use of phytoextraction to shallow soils (Pivetz, 2001).

Phytoextraction initially focused on hyperaccumulator plants, plants that accumulate a particular metal from metal contaminated soil to a very high degree (such as 100-fold or 1000-fold) when compared with other plants in that soil. These plants may reach some unusually high concentrations of metal in some of its tissues. These plants are relatively rare and found only in restricted areas around the world, with less than four hundred identified species for eight metals (Pivetz, 2001). But metals can be taken up by other plants that do not accumulate as high metal concentrations as hyperaccumulator plants, for example, corn (*Zea may*), sorghum (*Sorghum bicolor*), alfalfa (*Medicago sativa L.*) and willow trees (*Salix spp.*) (Pivetz, 2001) or sunflowers (Rizwan *et al.* 2016), namely plants that produce high amounts of biomass. The larger biomass of these plants could result in a higher amount of metals being removed from the soil even though metal concentrations within the plants might be lower than in hyperaccumulator plants (Pivetz, 2001).

An important question that arises after using plants for phytoextraction of metals from contaminated soils is: what will be the fate of the plant biomass? In fact, in recent years, the disposal of plants biomass used in phytoremediation has gained a lot of attention. Direct dumping, a stack of decay heat, burning, high-temperature decomposition and chemical extraction have all been suggested (Cao *et al.*, 2015, and references therein). In addition, some economic opportunities exist for plant biomass after being used in phytoextraction. For instance, plant biomass can be treated for recovery of precious and semiprecious metals, so-called phytomining (Cao *et al.*, 2015, and references therein). Furthermore, fast-growing and high-biomass producing plants, such as willow and poplar, could be used for both phytoremediation and bioenergy production (Abhilash *et al.*, 2012). Plant biomass can be used for different energy-recovery techniques, such as anaerobic digestion, incineration, gasification and production of biodiesel (Tian & Zhang, 2016, and reference therein). Some economic balances have showed that this strategy can produce some economic gains (Tian & Zhang, 2016, and reference there in). However, some questions concerning the production of bioenergy from phytoremediation residues are still unclear (Tian & Zhang, 2016). For instance, harvested metal-contaminated shoots can be introduced in the anaerobic digestor as biomass source. But it is important to be sure that the metal burden, namely of toxic metals such as Cd, in plant biomass will not affect biogas production (Tian & Zhang 2016). So, it is essential to assess these effects of metals concentrations on digestion systems and the design of anaerobic digestion processes (Cao *et al.*, 2015, and reference therein). Moreover, selecting suitable plants is essential, as species accumulating high concentrations of pollution may be difficult to digest (Tian & Zhang 2016).

To conclude, despite the fact that re-use of biomass used for metals phytoextraction is still not universally accepted, several advantages on the integration of phytoremediation technology with bioenergy production are already known. And other avenues also need to be explored. For instance, this type of plant biomass could be used to supplement the metals needed in an anaerobic digestor, contributing to the implementation of circular economic strategies and closing the loop. So, more research on this topic is required to promote an efficient application of phytoremediation.

7.4 OPPORTUNITIES FOR NEW USES OF DIGESTATE

Anaerobic digester (AD) developers are continuously faced with falling rates for renewable energy, the loss of the state investment credit grants and increasing costs of maintaining and operating the facilities. So, the financial models have to be changed and non-energy revenue streams for AD facilities are urgent in order to help protect feasibility and enhance the technology's already substantial environmental benefits. Since the power rate is flat, additional income streams have to be found to cover increased costs, especially in the regions where electricity and natural gas prices are low (e.g., North America). Digestate has already been classified as a product of lower value which generates minimal income, but demand for organic fertilisers and nutrient management can be a starter in development of the techniques and technology for digestate valorisation (WRAP, 2012) in order to: (i) increase the value of the digestate; (ii) create new markets for digestate products; (iii) reduce the dependence on land application; (iv) ensure more secure and sustainable outlets for digestate products; and (v) potentially reduce the operating cost of the facility.

Apart from the traditional land applications, digestate can be validated by: (i) the use of the digestate liquor for replacing freshwater and nutrients in algae cultivation; (ii) the use of solid digestate for energy production through biological (i.e., AD, bioethanol) or thermal processes (i.e., combustion, hydrothermal carbonization and pyrolysis); or (iii) the conversion of solid digestate into added-value products (char or activated carbons) through a pyrolysis process (Monlau *et al.*, 2015).

7.4.1 Land application

Traditionally, the solid fraction of digestate is often used as a soil fertiliser, or dried for the use in animal bedding, while the liquid fraction is usually spread on the fields. The focus remains on the products that will enhance soil properties as fertiliser and soil conditioner on farms, grass courts and home gardens. The nutrients (large parts nitrogen and potassium in liquid portion, phosphorous in the solid fraction; Liedl *et al.*, 2006) are separated and concentrated to create organic fertilisers in liquid or dry format. The digestate originating from manure or cellulosic wastes is loaded with absorbent fibres and mainly used in products that improve the soil's ability to control moisture and nutrient release (Gorrie, 2014).

Separation and concentration of the nutrients to create organic fertilisers in liquid or dry format can be a way of converting valuable digestate ingredients (nutrients and fibres) into co-products that will generate revenue. Recovered nutrients from digestates are applied either as a fertiliser or as a base feedstock for fertiliser production. Ammonium and phosphorus can be extracted from the digestate by precipitation in the form of magnesium ammonium phosphate (struvite) for use as an inorganic fertiliser or a feedstock for fertiliser production. By using a number of different commercially available techniques, ammonia, in the form of ammonium sulphate and ammonium nitrate, can also be recovered from the digestate. (WRAP, 2012.)

The digestate can either be composted on its own or co-composted with a range of standard composting feedstock, such as wood chip and green waste (Zeng *et al.*, 2016). Co-composting is beneficial for both waste streams because digestate provides a source of nitrogen, phosphorus, magnesium and iron, as well as moisture; and the standard composting feedstock provides a bulking agent, improving the carbon/nitrogen ratio and consistency of the final product (Evans, 2008).

Another possibility to give new value to the manure fibres is making a blend of nutrient-rich digested manure and other recycled natural materials for use in organic production, as a peat moss alternative. Peat moss is used in the burgeoning business of home gardening, but significant greenhouse gas emissions (both carbon and methane) are associated with the harvesting (for every acre harvested, 2400 tons of methane are released). An example for successful market utilisation of digested dairy manure fibres and improved long-term success of facilities is a bagged potting soil product (named 'Magic Dirt') by Cenergy, SA which has the ability to hold more than three times its weight in water and makes this blend suitable for use as a peat alternative (Goldstein, 2014).

7.4.2 Algal treatment of digestate

Algal treatment of digestate is an innovative approach for enhancement of the digestate liquid fraction. Liquid digestate can be combined with carbon dioxide to proliferate some microalgae until they can be harvested and used for the production of biochemicals and biofuels. Because of their easy production, growth rate, short lifecycles and independence from fertile agricultural land (land requirement for microalgae cultivation is estimated at 3% of traditional direct land application of digestate; Xia & Murphy, 2016), algae have great potential for energy use compared with conventional plants. Microalgae can fully use nutrients from liquid digestate, and CO_2 that is otherwise emitted to the atmosphere. The results of some recent works demonstrate the possibility of improving biomass accumulation (Xia & Murphy, 2016) and/or lipid production (Zuliani *et al.*, 2016) using different anaerobic digestates.

7.4.3 Bioenergy production

7.4.3.1 Bioethanol production

Valorisation of both solid and liquid digestate fractions can be achieved through biological fermentation and bioethanol production. The solid fraction has recently attracted attention for bioethanol production due to its high content of cellulose fibres (Xia & Murphy, 2016). In order to solubilise lignin that can limit carbohydrate availability and increase the cellulose content, treatments (mostly dilute-alkali treatment) have to be applied prior to enzymatic hydrolysis and fermentation.

Digestate shows several advantages for bioethanol production (Xia & Murphy, 2016): (i) enriched in easily accessible cellulose; (ii) better enzymatic digestibility than raw material (contains less hemicelluloses and more cellulose); (iii) AD process could improve the energy efficiency in traditional bioethanol production (reduce the energy requirement for the biomass milling).

The liquid digestate can also be used as a culture medium to replace freshwater and nutrients in bioethanol production process. Compounds, such as nutrients (N, P and K) and minerals (Mg, Zn, Cu) are essential for the enzymatic activity and yeast growth. Besides that, liquid digestate contains reduced amounts of potentially inhibitory compounds (i.e., furans and phenolics) for ethanol fermentation, since those compounds can be degraded in the AD process.

7.4.3.2 Thermal processes

Thermal digestate applications use heat (via incineration, combustion, hydrothermal carbonization or pyrolysis) in order to recover energy from the digestate and improve the overall energy efficiency of AD processes. Incineration is applicable for digestates with a high calorific value or where land-based application is not financially practical. Combustion is a thermochemical process with the complete oxidation of organic wastes to heat energy. The calorific value of digestate pellets was found to be similar to the calorific value of wood: 16.5 and 17.3 MJ kg^{-1} DM, respectively (Kratzeisen *et al.*, 2010). The residual ash can be used as a construction material for roads or for concrete production, and phosphorus can be recovered from the ash by acid leaching (WRAP, 2012).

Hydrothermal carbonization is a technique where wet organic material is converted into carbon-rich products called "hydrochars" with physicochemical properties close to fossil coal (Hoffmann *et al.*, 2013). Within the pyrolysis process, the digestate is heated under an oxygen-free atmosphere, producing biochar and vapour. By cooling the vapour phase, liquid is condensed (bio-oil composed of a large range of compounds including mainly sugars, acids, ketones, phenols and furans compounds) and the remaining gas phase (syngas) consists of mainly hydrogen, methane and carbon monoxide (Wang *et al.*, 2014). Incineration, pyrolysis and combustion require a solid digestate with a low moisture content, and thus a severe drying pre-treatment, so using heat from CHP facilities for efficient operation.

7.4.3.3 Bioelectricity

Microbial fuel cells are an application of fuel cells with the potential to remove nutrients and produce bioelectricity by the biological oxidation of organic matter from different wastes, as well as anaerobic digestate (Di Domenico *et al.*, 2015). The reactions take place under anaerobic conditions, and currently, this process is operational only at laboratory and pilot scales.

7.5 GAPS AND CHALLENGES

Trace element fluxes in the soils have been widely studied by phytomanagement. Research has shown that over time and under specific environmental conditions, but not in all studied cases, trace elements accumulate in soils. It is important to make the distinction between trace elements that interact with plants (phytoavailable) and trace elements that interact with other organisms (bioavailable). Physical contact between trace elements and plant roots is necessary for phytoavailability and affects root growth and plant uptake. High concentrations of trace elements in soils can damage plant tissues (oxidative stress) and hinder essential nutrients paths (Robinson *et al.*, 2009).

There is currently no general rule on whether ongoing digestate applications will result in constantly improved soil functions in the long run. Therefore, the measurement for assessing the health of a soil by determining diversity of the soil microbial community on structural and functional levels is essential for the sustainable production of crops and stability of arable land ecosystems (Riding *et al.*, 2015).

One of the major limitations to current research is the point-to-point exploration of strategies with no unified strategy (Murovec *et al.*, 2015). Using a diluted approach in which each the researcher can use his or her own approaches to studying the problem at hand, with little or no overlap from other researchers, has limited potential to unravel the cross correlations between various parts of the system and hence results in systematically contradicting results.

There is a chain of system characteristics that govern decision-making policy of digestate disposal: (i) geological characteristics of soils underpinning the production of plethora of biogas substrates; (ii) the substrate mixtures and ratios used, WWS characteristics, annual variations; (iii) anaerobic process characteristics; (iv) all of the previous points pivoting on digestate characteristics that affects the projected area of disposal; and finally (v) returning again to the geological characteristics of soils, which might be well different from those from which the original substrates were either grown or farmed or chemically produced.

Anaerobic digestion and the fate of its concentrated end products are characterized by a complex interplay of multiple factors acting over multiple scales, for example, landscape to nano-scale of metal environmental-matrix interactions. This is an emerging interdisciplinary framework that aims to improve our understanding, prevention of undesired environmental effects, and improve the use of crucial nutrients by integrating knowledge and data across multiple levels of life sciences (chemical, agricultural, engineering, microbiolgical, biotechnological and medical). As this is a multiscale system, the ultimate challenge and vision is a radical paradigm shift from a plethora of scale-specific reductionistic approaches to a more unified multiscale anaerobic systems science integrating past lessons and synchronizing research approaches, similar to current initiatives in medicine and physics.

REFERENCES

Abhilash P. C., Powell J. R., Singh H. B. and Singh B. K. (2012). Plant–microbe interactions: novel applications for exploitation in multipurpose remediation technologies. *Trends in Biotechnology*, **30**, 416–420.

Ahmad M., Rajapaksha A. U., Lim J. E., Zhang M., Bolan N., Mohan D., Vithanage M., Lee S. S. and Ok Y. S. (2014). Biochar as a sorbent for contaminant management in soil and water: a review. *Chemosphere*, **99**, 19–33.

Alburquerque J. A., de la Fuente C., Campoy M., Carrasco L., Nájera I., Baixauli C., Caravaca F., Roldán A., Cegarra J. and Bernal M. P. (2012). Agricultural use of digestate for horticultural crop production and improvement of soil properties. *European Journal of Agronomy*, **43**, 119–128.

Ali H., Khan E. and Sajad M. A. (2013). Phytoremediation of heavy metals—Concepts and applications. *Chemosphere*, **91**, 869–881.

Al Seadi T. and Lukehurst C. (2012). Quality Management of Digestate from Biogas Plants Used as Fertilizer. IEA Bioenergy (Technical Brochure).

Bauer A., Hervig M., Hopfer-Sixt K. and Amon T. (2009). Detailed monitoring of two biogas plants and mechanical solid-liquid separation of fermentation residues. *Journal of Biotechnology*, **142**, 56–63.

Bekiaris G., Peltre C., Jensen L. S. and Bruun S. (2016). Using FTIR-photoacoustic spectroscopy for phosphorus speciation analysis of biochars. *Spectrochimica Acta Part A: Molecular and Biomolecular Spectroscopy*, **168**, 29–36.

Braun R. and Wellinger A. (2003). 'Potential of Codigestion', www.iea-biogas.ne (accessed 25 January 2010).

Bruun S., Harmer S. L., Bekiaris G., Christel W., Zuin L., Hu Y., Jensen L. S. and Lombi E. (2017). The effect of different pyrolysis temperatures on the speciation and availability in soil of P in biochar produced from the solid fraction of manure. *Chemosphere*, **169**, 377–386.

Cao Z., Wang S., Wang T., Chang Z., Shen Z. and Chen Y. (2015). Using Contaminated Plants Involved in Phytoremediation for Anaerobic Digestion. *International Journal of Phytoremediation*, **17**, 201–207.

Capson-Tojo G., Moscoviz R., Ruiz D., Santa-Catalina G., Trably E., Rouez M., Crest M., Steyer J. P., Bernet N., Delgenès J. P. and Escudié R. (2018). Addition of granular activated carbon and trace elements to favor volatile fatty acid consumption during anaerobic digestion of food waste. *Bioresource Technology*, **260**, 157–168.

Capson-Tojo G., Girard C., Rouez M., Crest M., Steyer J. P., Bernet N., Delgenès J. P. and Escudié R. (2019). Addition of biochar and trace elements in the form of industrial FeCl3 to stabilize anaerobic digestion of food waste: dosage optimization and long-term study. *Journal of Chemical Technology & Biotechnology*, **94**, 505–515.

Čermák P., Kohel J. and Dedera F. (2002). Reclamation of areas devastated by mining activities in the North Bohemian lignite basin (methodology for practice). Monograph, 88 p., first edition. Research Institute for Soil and Water Conservation. Prague. (In Czech).

CEU (Council of the European Union) (1991a). European Parliament and Council directive of 21 May 1991, concerning urban waste-water treatment (1991/271/EC). *Official Journal of the European Union L*, **135**, 40–52.

CEU (1991b). European Parliament and of the Council directive of 12 December 1991, concerning the protection of waters against pollution caused by nitrates from agricultural sources (1991/676/EC). *Official Journal of the European Union LL*, **375**, 1–8.

CEU (1999). European Parliament and of the Council directive of 26 April 1999, on the landfill of waste (1999/31/EC). *Official Journal of the European Union L*, **182**, 1–19.

CEU (2000). European Parliament and of the Council directive of 4 December 2000, on the incineration of waste (2000/76/EC). *Official Journal of the European Union LL*, **332**, 91–111.

CEU (2008). European Parliament and Council directive of 19 November 2008, on waste and repealing certain (2008/98/EC). *Official Journal of the European Union LL*, **312**, 3–30.

Chantigny M. H., Angers D. A., Rochette P., Bélanger G., Massé D. and Côté D. (2007). Gaseous nitrogen emissions and forage nitrogen uptake on soils fertilized with raw and treated swine manure. *Journal of Environmental Quality*, **36**, 1864–1872. https://doi.org/10.2134/jeq2007.0083.

Dahlin J., Herbes C. and Nelles M. (2015). Biogas digestate marketing: qualitative insights into the supply side. *Resources, Conservation and Recycling*, **104**, 152–161.

Diacono M. and Montemurro F. (2010). Long-term effects of organic amendments on soil fertility. A review. *Agronomy for Sustainable Development*, **30**, 401–422.

Di Domenico E., Petroni G., Mancini D., Geri A., Di Palma L. and Ascenzioni F. (2015). Development of electroactive and anaerobic ammonium-oxidizing (Anammox) biofilms from digestate in microbial fuel cells. *BioMed Research International*. Article ID 351014, 10 pages. https://www.hindawi.com/journals/bmri/2015/351014/cta/.

Drosg B., Fuchs W., Al Seadi T., Madsen M. and Linke B. (2015). Nutrient Recovery by Biogas Digestate Processing. IEA Bioenergy, Task 37. ISBN 978-1-910154-16-8 Available from: https://www.ieabioenergy.com/publications

EC (2003). Regulation (EC/2003/2003). of the European parliament and of the council of 13 October 2003 relating to fertilisers. *Official Journal of the European Union L*, **182**, 1–193.

EC (2007). Council Regulation (EC/834/2007). of 28 June 2007 on organic production and labelling of organic products and repealing Regulation (EEC/ 2092/91). *Official Journal of the European Union L*, **189**, 1–23.

EC (2009). Regulation (EC/1069/2009). of the European parliament and of the council of 21 October 2009 laying down health rules as regards animal by-products and derived products not intended for human consumption and repealing Regulation (EC) No 1774/2002 (Animal by-products Regulation). *Official Journal of the European Union L*, **300**, 1–32.

Eibisch N., Schroll R., Fuß R., Mikutta R., Helfrich M. and Flessa H. (2015). Pyrochars and hydrochars differently alter the sorption of the herbicide isoproturon in an agricultural soil. *Chemosphere*, **119**, 155–162.

Evans T. D. (2008). An Independent Review of Sludge Treatment Processes and Innovations. Australian Water Association Biosolids Conference, Adelaide, Australia.

Fagbohungbe M. O., Herbert B. M., Hurst L., Ibeto C. N., Li H., Usmani S. Q. and Semple K. T. (2017). The challenges of anaerobic digestion and the role of biochar in optimizing anaerobic digestion. *Waste Management*, **61**, 236–249.

Feng Q. and Lin Y. (2017). Integrated processes of anaerobic digestion and pyrolysis for higher bioenergy recovery from lignocellulosic biomass: a brief review. *Renewable and Sustainable Energy Reviews*, **77**, 1272–1287.

Fu D., Chen Z., Xia D., Shen L., Wang Y. and Li Q. (2017). A novel solid digestate-derived biochar-Cu NP composite activating H_2O_2 system for simultaneous adsorption and degradation of tetracycline. *Environmental Pollution*, **221**, 301–310.

Gerardi M. H. (2003). The Microbiology of Anaerobic Digesters. John Wiley & Sons, London, UK.

Goldstein N. (2014). Digested Dairy Manure To High-End Potting Soil. *BioCycle*, **55**, 48.

Gómez X., Cuetos M. J., García A. I. and Morán A. (2007). An evaluation of stability by thermogravimetric analysis of digestate obtained from different biowastes. *Journal of Hazardous Materials*, **149**, 97–105.

Gorrie P. (2014). Capitalizing On Digester Coproducts. *BioCycle*, **55**, 48.

Gusiatin Z. M., Kurkowski R., Brym S. and Wiśniewski D. (2016). Properties of biochars from conventional and alternative feedstocks and their suitability for metal immobilization in industrial soil. *Environmental Science and Pollution Research*, **23**, 21249–21261.

Hargreaves J. C., Adl M. S. and Warman P. R. (2008). A review of the use of composted municipal solid waste in agriculture. *Agriculture, Ecosystems & Environment*, **123**, 1–14.

Himanen M. and Hänninen K. (2011). Composting of bio-waste, aerobic and anaerobic sludges – effect of feedstock on the process and quality of compost. *Bioresource Technology*, **102**, 2842–2852.

Hoffmann J., Rudra S., Toor S. S., Holm-Nielsen J. B. and Rosendahl L. A. (2013). Conceptual design of an integrated hydrothermal liquefaction and biogas plant for sustainable bioenergy production. *Bioresource Technology*, **129**, 402–410.

Holm-Nielsen J., Halberg N., Hutingford S. and Al Seadi T. (1997). Joint Biogas Plants, Agricultural Advantages-Circulation of N, P, and K. Report, second edition, August 1997, Danish Energy Agency, Copenhagen, Denmark. https://static1.squarespace.com/static/56d1a94e1bbee09a4bc6c578/t/56d4ba502b8ddea29eb2872f/1456781906892/PAS110_2014_final.pdf.

Hung C. Y., Tsai W. T., Chen J. W., Lin Y. Q. and Chang Y. M. (2017). Characterization of biochar prepared from biogas digestate. *Waste Management*, **66**, 53–60.

Igoni A. H., Ayotamuno M. J., Eze C. L., Ogaji S. O. T. and Probert S. D. (2008). Designs of anaerobic digesters for producing biogas from municipal solid-waste. *Applied Energy*, **85**, 430–438.

Insam H., Gómez-Brandón M. and Ascher J. (2015). Manure-based biogas fermentation residues – Friend or foe of soil fertility?. *Soil Biology and Biochemistry*, **84**, 1–14. https://doi.org/http://dx.doi.org/10.1016/j.soilbio.2015.02.006.

Inyang M. and Dickenson E. (2015). The potential role of biochar in the removal of organic and microbial contaminants from potable and reuse water: a review. *Chemosphere*, **134**, 232–240.

Jiang B., Lin Y. and Mbog J. C. (2018). Biochar derived from swine manure digestate and applied on the removals of heavy metals and antibiotics. *Bioresource Technology*, **270**, 603–611.

Kataki S., Hazarika S. and Baruah D. C. (2017). Assessment of by-products of bioenergy systems (anaerobic digestion and gasification) as potential crop nutrient. *Waste Management*, **59**, 102–117, doi:10.1016/j.wasman.2016.10.018.

Kirchmann H. and Witter E. (1992). Composition of fresh, aerobic and anaerobic farm animal dungs. *Bioresource Technology*, **40**, 137–142.

Kratzeisen M., Starcevic N., Martinov M., Maurer C. and Muller J. (2010). Applicability of biogas digestate as solid fuel. *Fuel*, **89**, 2544–2548.

Laird D. A. (2008). The charcoal vision: a win–win–win scenario for simultaneously producing bioenergy, permanently sequestering carbon, while improving soil and water quality. *Agronomy Journal*, **100**, 178–181.

Laureni M., Palatsi J., Llovera M. and Bonmat A. (2013). Influence of pig slurry characteristics on ammonia stripping efficiencies and quality of the recovered ammonium-sulfate solution. *Journal of Chemical Technology and Biotechnology*, **88**, 1654–1662.

Ledda C., Schievano A., Salati S. and Adani F. (2013). Nitrogen and water recovery from animal slurries by a new integrated ultrafiltration, reverse osmosis and cold stripping process: a case study. *Water Research*, **47**, 6157–6166.

Lehmann J. and Joseph S. (eds) (2015). Biochar for Environmental Management: Science, Technology and Implementation. Routledge.

Liedl B. E., Bombardiere J. and Chatfield J. M. (2006). Fertiliser potential of liquid and solid effluent from thermophilic anaerobic digestion of poultry waste. *Water Science and Technology*, **53**(8), 69–79.

Liu H. (2016). Achilles heel of environmental risk from recycling of sludge to soil as amendment: a summary in recent ten years (2007–2016). *Waste Management*, **56**, 575–583. https://doi.org/http://dx.doi.org/10.1016/j.wasman.2016.05.028.

Lü F., Hua Z., Shao L. and He P. (2018). Loop bioenergy production and carbon sequestration of polymeric waste by integrating biochemical and thermochemical conversion processes: a conceptual framework and recent advances. *Renewable Energy*, **124**, 202–211.

Lukehurst C. T., Frost P. and Al Seadi T. (2010). Utilisation of Digestate from Biogas Plants as Biofertiliser. IEA Bioenergy, Paris.

Luz F. C., Cordiner S., Manni A., Mulone V. and Rocco V. (2018). Biochar characteristics and early applications in anaerobic digestion-a review. *Journal of Environmental Chemical Engineering*, **6**, 2892–2909.

Ma Y., Oliveira R. S., Freitas H. and Zhang C. (2016). Biochemical and molecular mechanisms of plant-microbe-metal interactions: relevance for phytoremediation. *Frontiers in Plant Science*, **7**, Article 918 (19 pages).

Makádi M., Tomócsik A. and Orosz V. (2012). Digestate: a new nutrient source – review. In: S. Kumar (ed.), Biogas. InTech Europe, Croatia, pp. 295–310.

Masciandaro G. and Ceccanti B. (1999). Assessing soil quality in different agro-ecosystems through biochemical and chemico-structural properties of humic substances. *Soil & Tillage Research*, **51**, 129–137.

Mata-Alvarez J., Dosta J., Romero-Güiza M. S., Fonoll X., Peces M. and Astals S. (2014). A critical review on anaerobic co-digestion achievements between 2010 and 2013. *Renewable and Sustainable Energy Reviews*, **36**, 412–427.

Moestedt J., Malmborg J. and Nordell E. (2015). Determination of methane and carbon dioxide formation rate constants for semi-continuously fed anaerobic digesters. *Energies*, **8**, 645–655.

Mohan D., Sarswat A., Ok Y. S. and Pittman C. U. Jr., (2014). Organic and inorganic contaminants removal from water with biochar, a renewable, low cost and sustainable adsorbent–a critical review. *Bioresource Technology*, **160**, 191–202.

Møller H. B., Lund I. and Sommer S. G. (2000). Solid-liquid separation of livestock slurry: efficiency and cost. *Bioresource Technology*, **74**, 223–229.

Møller K. (2015). Effects of anaerobic digestion on soil carbon and nitrogen turnover, N emissions, and soil biological activity. *A review. Agronomy for Sustainable Development*, **35**, 1021–1041.

Möller K. and Müller T. (2012). Effects of anaerobic digestion on digestate nutrient availability and crop growth: A review: Digestate nutrient availability. *Engineering in Life Sciences*, **12**, 242–257, doi:10.1002/elsc.201100085.

Möller K. and Stinner W. (2009). Effects of different manuring systems with and without biogas digestion on soil mineral nitrogen content and on gaseous nitrogen losses (ammonia, nitrous oxides). *European Journal of Agronomy*, 30, 1–16. https://doi.org/10.1016/j.eja.2008.06.003

Monlau F., Sambusiti C., Ficara E., Aboulkas A., Barakata A. and Carrère H. (2015). New opportunities for agricultural digestate valorization: current situation and perspectives. *Energy and Environmental Science*, **8**, 2600–2621.

Monlau F., Francavilla M., Sambusiti C., Antoniou N., Solhy A., Libutti A., Zabaniotou A., Barakat A. and Monteleone M. (2016). Toward a functional integration of anaerobic digestion and pyrolysis for a sustainable resource management. *Comparison between Solid-digestate and Its Derived Pyrochar as Soil Amendment. Applied Energy*, **169**, 652–662.

Mumme J., Eckervogt L., Pielert J., Diakité M., Rupp F. and Kern J. (2011). Hydrothermal carbonization of anaerobically digested maize silage. *Bioresource Technology*, **102**, 9255–9260.

Mumme J., Srocke F., Heeg K. and Werner M. (2014). Use of biochars in anaerobic digestion. *Bioresource Technology*, **164**, 189–197.

Murovec B., Kolbl S. and Stres B. (2015). Methane Yield Database: Online infrastructure and bioresource for methane yield data and related metadata. *Bioresource Technology*, **189**, 217–223.

Nansubuga I., Banadda N., Ronsse F., Verstraete W. and Rabaey K. (2015). Digestion of high rate activated sludge coupled to biochar formation for soil improvement in the tropics.

Water Research, **81**, 216–222.

Nkoa R. (2014). Agricultural benefits and environmental risks of soil fertilization with anaerobic digestates: a review. *Agronomy for Sustainable Development*, **32**, 473–492. https://doi.org/10.1007/s13593-013-0196-z.

Odlare M., Pell M. and Svensson K. (2008). Changes in soil chemical and microbiological properties during 4 years of application of various organic residues. *Waste Management (New York N. Y.)*, **28**, 1246–1253, https://doi.org/10.1016/j.wasman. 2007.06.005.

Oliveira T., Mucha A. P., Reis I., Rodrigues P., Gomes C. R. and Almeida C. M. R. (2014). Copper phytoremediation by a salt marsh plant (Phragmites australis) enhanced by autochthonous bioaugmentation. *Marine Pollution Bulletin*, **88**, 231–238.

Pawlett M. and Tibbett M. (2015). Is sodium in anaerobically digested food waste a potential risk to soils? *Sustainable Environment Research*, **25**, 235–239.

Pivetz B. E. (2001). Phytoremediation of Contaminated Soil and Ground Water At Hazardous Waste Sites, Ground Water Issue. National Risk Management Research Laboratory Subsurface Protection and Remediation Division Robert S. Kerr Environmental Research Center Ada, Oklahoma, Technology Innovation Office, Office of Solid Waste and Emergency Response, US EPA, Washington, DC, USA.

Pognani M., D'Imporzano G., Scaglia B. and Adani F. (2009). Substituting energy crops with organic fraction of municipal solid waste for biogas production at farm level: A full-scale plant study. *Process Biochemistry*, **44**, 817–821.

Pöschl M., Ward S. and Owende P. (2010). Evaluation of energy efficiency of various biogas production and utilization pathways. *Applied Energy*, **87**, 3305–3321.

Rajagopal R., Masse D. I. and Singh G. (2013). A critical review on inhibition of anaerobic digestion process by excess ammonia. *Bioresource Technology*, **143**, 632–641.

Rajkumar M., Sandhya S., Prasad M. N. V. and Freitas H. (2012). Perspectives of plant-associated microbes in heavy metal phytoremediation. *Biotechnological Advances*, **30**, 1562–1574.

Riding M. J., Herbert B. M. J., Ricketts L., Dodd I., Ostle N. and Semple K. T. (2015). Harmonising conflicts between science, regulation, perception and environmental impact: The case of soil conditioners from bioenergy. *Environment International*, **75**, 52–67, https://doi.org/http://dx.doi.org/10.1016/j.envint.2014.10.025.

Rizwan M., Ali S., Rizvi H., Rinklebe J., Tsang D. C. W., Meers E., Ok Y. S. and Ishaque W. (2016). Phytomanagement of heavy metals in contaminated soils using sunflower: a review. *Critical Reviews in Environmental Science and Technology*, **46**, 1498–1528.

Robinson B. H., Bañuelos G., Conesa H. M., Evangelou M. W. H. and Schulin R. (2009). The Phytomanagement of Trace Elements in Soil. *Critical Reviews in Plant Sciences*, **28**, 240–266, https://doi.org/10.1080/07352680903035424.

Rocha C., Almeida C. M. R., Basto M. C. P. and Vasconcelos M. T. S. D. (2014). Antioxidant response of *Phragmites australis* to Cu and Cd contamination. *Ecotoxicology and Environmental Safety*, **109**, 152–160.

Rocha C., Almeida C. M. R., Basto M. C. P. and Vasconcelos M. T. S. D. (2016). Marsh plants response to metals: exudation of aliphatic low molecular weight organic acids (ALMWOAs). *Estuarine Coastal and Shelf Science*, **171**, 77–84.

Roschke M. and Plöchl M. (2006). Eingeschaften und Zusammensetzung der Gärreste. Postdam: Biogas in der Landwirtschaft. Litfaden für Landwirte und Investoren im Land Brandenburg. (In German).

Schouten S., van Groenigen J. W., Oenema O. and Cayuela M. L. (2012). Bioenergy from cattle manure? Implications of anaerobic digestion and subsequent pyrolysis for carbon and nitrogen dynamics in soil. *GCB Bioenergy*, **4**, 751–760.

Silva M. N., Mucha A. P., Rocha A. C., Teixeira C., Gomes C. R. and Almeida C. M. R. (2014). A strategy to potentiate Cd phytoremediation by saltmarsh plants – autochthonous bioaugmentation. *Journal of Environmental Management*, **134**, 136–144.

Silvestri S., Hansson A., Thorpe J., Pomykala R. (2013). Four detailed strategies for residuals product management, Report within project "Biomethane as an Alternative Source for Transport and Energy Renaissance (BIOMASTER)" (No. IEE/10/351/SI2.591136).

Sommer S. G. and Husted S. (1995). The chemical buffer system in raw and digested animal slurry. *Journal of Agricultural Science*, **124**, 45e53.

Stefaniuk M. and Oleszczuk P. (2015). Characterization of biochars produced from residues from biogas production. *Journal of Analytical and Applied Pyrolysis*, **115**, 157–165.

Steffen R., Szolar O. and Braun R. (1998). Feedstocks for Anaerobic Digestion. University of Agricultural Science, Vienna, Austria.

Stinner W., Möller K. and Leithold G. (2008). Effect of biogas digestion of clover/grass-leys, cover crops and crop residues on nitrogen cycle and crop yield in organic stockless farming system. *European Journal of Agronomy*, **29**, 125–134.

Takaya C. A., Fletcher L. A., Singh S., Anyikude K. U. and Ross A. B. (2016). Phosphate and ammonium sorption capacity of biochar and hydrochar from different wastes. *Chemosphere*, **145**, 518–527.

Tambone F., Genevini P., D'Imporzano G. and Adani F. (2009). Assessing amendment properties of digestate by studying the organic matter composition and the degree of biological stability during the anaerobic digestion of the organic fraction of MSW. *Bioresource Technology*, **100**, 3140–3142.

Tambone F., Terruzzi L., Scaglia B. and Adani F. (2015). Composting of the solid fraction of digestate derived from pig slurry: Biological processes and compost properties. *Waste Management*, **35**, 55–61.

Teglia C., Tremier A. and Martel J.-L. (2011). Characterization of solid digestates: Part 1. Review of existing indicators to assess solid digestates agricultural use. *Waste and Biomass Valorization*, **2**, 43–58, https://doi.org/10.1007/s12649–010–9051–5.

Tian Y. and Zhang H. (2016). Producing biogas from agricultural residues generated during phytoremediation process: Possibility, threshold, and challenges. *International Journal of Green Energy*, **13**, 1556–1563.

Tombácz E., Szekeres M., Baranyi L. and Micheli E. (1998). Surface modification of clay minerals by organic polyions. *Colloids and Surfaces A*, **141**, 379–384.

Tombácz E., Filipcsei G., Szekeres M. and Gingl Z. (1999). Particle aggregation in complex aquatic systems. *Colloids and Surfaces A*, **151**, 233–244.

Usťak S., Püschel D., Rydlová J., Gryndler M., Mikanová O. and Vosátka M. (2010). Cultivation of selected non-food crops in combination with applications of organic fertilisers and microbiological preparations as a means of biological reclamation of anthropogenic soils (methodology for practice), ISBN 978-80-7427-053-6, Envi-Bio, 36 p. (In Czech).

Vaneeckhaute C., Meers E., Michels E., Ghekiere G., Accoe F. and Tack F. M. G. (2013). Closing the nutrient cycle by using bio-digestion waste derivatives as synthetic fertiliser substitutes: A field experiment. *Biomass Bioenerg.* **55**, 175–189.

Wambugu, C., Rene, E. R., Vossenberg, J., Dupont, C. and van Hullebusch, E. D. (2019).

Role of biochar in anaerobic digestion based biorefinery for food waste. *Frontiers in Energy Research*, 7, 14.

Wang T., Ye X., Yin J., Lu Q., Zheng Z. and Dong C. (2014). Effects of biopretreatment on pyrolysis behaviors of corn stalk by methanogen. *Bioresource Technology*, **164**, 416–419.

Weiland P. (2010). Biogas production: current state and perspectives. *Applied Microbiology and Biotechnology*, **85**, 849–60.

Wongrod S., Simon S., Guibaud G., Lens P. N., Pechaud Y., Huguenot D. and van Hullebusch E. D. (2018a). Lead sorption by biochar produced from digestates: Consequences of chemical modification and washing. *Journal of Environmental Management*, **219**, 277–284.

Wongrod S., Simon S., van Hullebusch E. D., Lens P. N., and Guibaud G. (2018b). Changes of sewage sludge digestate-derived biochar properties after chemical treatments and influence on As (III and V) and Cd (II) sorption. *International biodeterioration & biodegradation*, **135**, 96–102.

Wongrod S., Simon S., van Hullebusch E. D., Lens P. N. and Guibaud, G. (2019). Assessing arsenic redox state evolution in solution and solid phase during As (III) sorption onto chemically-treated sewage sludge digestate biochars. *Bioresource Technology*, **275**, 232–238.

WRAP (2012). Enhancement and Treatment of Digestates from Anaerobic Digestion, Desk Top Study on Digestate Enhancement and Treatment. Pell Frischmann Consultants Ltd, London, UK.

Wu S., Ni P., Li J., Sun H., Wang Y., Luo H., Dach J. and Dong R. (2016). Integrated approach to sustain biogas production in anaerobic digestion of chicken manure under recycled utilization of liquid digestate: dynamics of ammonium accumulation and mitigation control. *Bioresource Technology*, **205**, 75–81.

Xia A. and Murphy J. (2016). Microalgal Cultivation in Treating Liquid Digestate from Biogas Systems. *Trends in Biotechnology*, **34**, 264–275.

Yan F., Schubert S. and Mengel K. (1996). Sol pH increase due to biological decarboxalation of organic anions. *Soil Biology & Biochemistry*, **28**, 611–624.

Yan Z. Y., Song Z. L., Li D., Yuan Y. X., Liu X. F. and Zheng T. (2015). The effects of initial substrate concentration, C/N ratio, and temperature on solid-state anaerobic digestion from composting rice straw. *Bioresource Technology*, **177**, 266–273.

Zeng Y., De Guardia A. and Dabert P. (2016). Improving composting as a post-treatment of anaerobic digestate. *Bioresource Technology*, **201**, 293–303.

Zhang J., Zhao W., Zhang H., Wang Z., Fan C. and Zang L. (2018). Recent achievement in enhancing anaerobic digestion with carbon-based functional materials. *Bioresource Technology*, **266**, 555–567.

Zheng X., Yang Z., Xu X., Dai M. and Guo R. (2018). Characterization and ammonia adsorption of biochar prepared from distillers' grains anaerobic digestion residue with different pyrolysis temperatures. *Journal of Chemical Technology & Biotechnology*, **93**, 198–206.

Zuliani L., Frison N., Jelic A., Fatone F., Bolzonella D. and Ballottari M. (2016). Microalgae cultivation on anaerobic digestate of municipal wastewater, sewage sludge and agro-waste. *International Journal of Molecular Science*, **17**, 1692.

Permissions

All chapters in this book were first published by IWA Publishing; hereby published with permission under the Creative Commons Attribution License or equivalent. Every chapter published in this book has been scrutinized by our experts. Their significance has been extensively debated. The topics covered herein carry significant information for a comprehensive understanding. They may even be implemented as practical applications or may be referred to as a beginning point for further studies.

The contributors of this book come from diverse backgrounds, making this book a truly international effort. We would like to thank all the contributing authors for lending their expertise to make the book truly unique. They have played a crucial role in the development of this book. Without their invaluable contributions this book wouldn't have been possible. They have made vital efforts to compile up to date information on the varied aspects of this subject to make this book a valuable addition to the collection of many professionals and students.

This book was conceptualized with the vision of imparting up-to-date and integrated information in this field. To ensure the same, a matchless editorial board was set up. Every individual on the board went through rigorous rounds of assessment to prove their worth. After which they invested a large part of their time researching and compiling the most relevant data for our readers.

The editorial board has been involved in producing this book since its inception. They have spent rigorous hours researching and exploring the diverse topics which have resulted in the successful publishing of this book. They have passed on their knowledge of decades through this book. To expedite this challenging task, the publisher supported the team at every step. A small team of assistant editors was also appointed to further simplify the editing procedure and attain best results for the readers.

Apart from the editorial board, the designing team has also invested a significant amount of their time in understanding the subject and creating the most relevant covers. They scrutinized every image to scout for the most suitable representation of the subject and create an appropriate cover for the book.

The publishing team has been an ardent support to the editorial, designing and production team. Their endless efforts to recruit the best for this project, has resulted in the accomplishment of this book. They are a veteran in the field of academics and their pool of knowledge is as vast as their experience in printing. Their expertise and guidance has proved useful at every step. Their uncompromising quality standards have made this book an exceptional effort. Their encouragement from time to time has been an inspiration for everyone.

The publisher and the editorial board hope that this book will prove to be a valuable piece of knowledge for students, practitioners and scholars across the globe.

Index